Introduction to Engineering Design

McGraw-Hill's *BEST*—Basic Engineering Series and Tools

Chapman, *Introduction to Fortran 90/95*
D'Orazio and Tan, *C Program Design for Engineers*
Eide et al., *Introduction to Engineering Problem Solving*
Eide et al., *Introduction to Engineering Design*
Eisenberg, *A Beginner's Guide to Technical Communication*
Gottfried, *Spreadsheet Tools for Engineers with Excel*
Mathsoft's Student Edition of Mathcad 6.0
Palm, *Introduction to MATLAB®*

INTRODUCTION TO ENGINEERING DESIGN

Arvid R. Eide

Roland D. Jenison

Lane H. Mashaw

Larry L. Northup

Iowa State University

WCB McGraw-Hill

Boston, Massachusetts   Burr Ridge, Illinois
Dubuque, Iowa   Madison, Wisconsin   New York, NewYork
San Francisco, California   St. Louis, Missouri

McGraw-Hill

A Division of The **McGraw·Hill** Companies

INTRODUCTION TO ENGINEERING DESIGN

The material in this book is taken from Edie, Jenison, Mashaw, and Northup: *Engineering Fundamentals and Problem Solving*.

This book is printed on acid-free paper.

2 3 4 5 7 8 9 0   DOC/DOC   9 0 9 8

ISBN 0-07-018922-6

Editorial director: *Kevin Kane*
Publisher: *Tom Casson*
Sponsoring editor: *Eric M. Munson*
Development Editor II: *Holly Stark*
Marketing manager: *John Wannemacher*
Senior project manager: *Jean Lou Hess*
Production supervisor: *Heather D. Burbridge*
Designer: *Larry J. Cope*
Compositor: *York Graphic Services, Inc.*
Typeface: *10/12 Century Schoolbook*
Printer: *R. R. Donnelley & Sons Company*

Library of Congress Cataloging-in-Publication Data

Introduction to engineer desing / Arvid R. Eide . . . [et al.].
     p. cm. — (McGraw-Hill's BEST—basic engineering series and tools)
  Includes index.
  ISBN 0-07-018922-6
  1. Engineering design.   I. Eide, Arvid R.   II. Series.
TA3300.I64   1998
620'.0042—dc21   97-8465

http://www.mhcollege.com

About the Authors

Arvid R. Eide is a native Iowan. He received his baccalaureate degree in mechanical engineering from Iowa State University in 1962. Upon graduation he spent two years in the U.S. Army as a commissioned officer and then returned to Iowa State as a full-time instructor while completing a master's degree in mechanical engineering. Professor Eide has worked for Western Electric, John Deere, and the Trane Company. He received his Ph.D. in 1974 and was appointed professor and chair of Freshman Engineering at Iowa State from 1974 to 1989. Dr. Eide was selected Associate Dean of Academic Affairs from 1989 to 1995 and currently serves as professor of Mechanical Engineering.

Roland D. Jenison is a professor of aerospace engineering and engineering mechanics at Iowa State University. He has 30 years of teaching experience in lower division engineering and engineering technology. He has taught courses in engineering design graphics and engineering problem solving and has published numerous papers in these areas. His scholarly activities include learning-based instruction in design graphics, design education, and design-build projects for lower division engineering students. He is a long-time member of ASEE and served as chair of the Engineering Design Graphics Division in 1986–1987.

Lane H. Mashaw earned his BSCE from the University of Illinois and MSCE from the University of Iowa. He served as a municipal engineer in Champaign, Ill., Rockford, Ill., and Iowa City, Ia., for nine years and then was in private practice in Decatur, Ill. for another nine years. He taught at the University of Iowa from 1964 to 1974 and at Iowa State University from 1974 until his retirement in 1987. He is currently emeritus professor of civil and construction engineering.

Larry L. Northup is a professor of civil and construction engineering at Iowa State University. He has more than 30 years of teaching experience, with the past 20 years devoted to lower division engineering courses in problem solving, graphics, and design. He has 2 years of industrial experience and is a registered engineer in Iowa. He has been active in ASEE (Engineering Design Graphics Division), having served as chair of the Freshman Year Committee and Director of Technical and Professional Committees (1981–1984). He also served as chair of the Freshman Programs Constituent Committee (now Division) of ASEE in 1983–1984.

Foreword

Engineering educators have had long-standing debates over the content of introductory freshman engineering courses. Some schools emphasize computer-based instruction, some focus on engineering analysis, some concentrate on graphics and visualization, while others emphasize hands-on design. Two things, however, appear certain: no two schools do exactly the same thing, and at most schools, the introductory engineering courses frequently change from one year to the next. In fact, the introductory engineering courses at many schools have become a smorgasbord of different topics, some classical and others closely tied to computer software applications. Given this diversity in content and purpose, the task of providing appropriate text material becomes problematic, since every instructor requires something different.

McGraw-Hill has responded to this challenge by creating a series of modularized textbooks for the topics covered in most first-year introductory engineering courses. Written by authors who are acknowledged authorities in their respective fields, the individual modules vary in length, in accordance with the time typically devoted to each subject. For example, modules on programming languages are written as introductory-level textbooks, providing material for an entire semester of study, whereas modules that cover shorter topics such as ethics and technical writing provide less material, as appropriate for a few weeks of instruction. Individual instructors can easily combine these modules to conform to their particular courses. Most modules include numerous problems and/or projects, and are suitable for use within an active-learning environment.

The goal of this series is to provide the educational community with text material that is timely, affordable, of high quality, and flexible in how it is used. We ask that you assist us in fulfilling this goal by letting us know how well we are serving your needs. We are particularly interested in knowing what, in your opinion, we have done well, and where we can make improvements or offer new modules.

Byron S. Gottfried
Consulting Editor
University of Pittsburgh

Preface

TO THE STUDENT

As you begin the study of engineering you are no doubt filled with enthusiasm, curiosity, and a desire to succeed. Your first year will be spent primarily establishing a solid foundation in mathematics, basic sciences, and communications. You may at times question what the benefits of this background material are and when actual engineering experiences will begin. We believe that they begin now. Additionally, we believe that the material presented in this module will provide you a fundamental understanding of how engineers approach design in today's technological world.

TO THE INSTRUCTOR

Engineering courses for first-year students continue to be in a state of transition. A diverse set of goals such as providing motivation to study engineering, introducing cooperative learning, and encouraging work in a team environment have each provided reasons to study engineering design during the first year. The traditional engineering drawing and descriptive geometry courses have been largely replaced with computer graphics and CAD-based courses. Courses in introductory engineering and problem solving are now utilizing spreadsheets and mathematical solvers in addition to teaching the rudiments of a computer language. The World Wide Web (WWW) is rapidly becoming a major instructional tool, providing a wealth of data to supplement class notes and textbooks.

Since 1974, students at Iowa State University have taken a course that has a major objective of introducing engineering design. This module has thus evolved from more than 20 years of experience with teaching engineering design to thousands of first-year students.

A 10-step design process is explained and supplemented with an actual preliminary design performed by a first-year student team. The process as described allows you to supplement the text material with personal examples to bring your own design experience into the classroom. Mathematical expertise beyond algebra, trigonometry and analytical geometry is not required for any material in the module.

ACKNOWLEDGMENTS

The authors are indebted to many persons who assisted in the development of this module. First we would like to thank the faculty of the former Division of Engineering Fundamentals and Multidisciplinary Design at Iowa State University who taught engineering design to freshman students over the past 20 years. They, with support of engineering faculty from other departments, have made the courses a success by their efforts. Several thousands of students have taken the courses, and we want to thank them for their comments and ideas that have influenced this module. The many suggestions of faculty and students alike have provided us with much information necessary to prepare this material. A special thanks to the reviewers whose suggestions were extremely valuable and greatly shaped the manuscript. We also express grateful appreciation to Jane Stowe who worked many hours to type the manuscript. Finally we thank our families for their constant support of our efforts.

Arvid R. Eide
Roland D. Jenison
Lane H. Mashaw
Larry L. Northup

Contents

The Engineering Profession

Introduction

The rapidly expanding sphere of science and technology may seem overwhelming to the individual seeking a career in a technological field. A technical specialist today may be called either engineer, scientist, technologist, or technician, depending upon education, industrial affiliation, or specific work. For example, nearly 350 colleges and universities offer engineering programs accredited by the Accreditation Board for Engineering and Technology (ABET) or the Canadian Engineering Accreditation Board (CEAB). Included are such traditional specialties as aerospace, agricultural, chemical, civil, electrical, industrial, and mechanical engineering—as well as the expanding areas of computer, energy, environmental, and materials engineering. Programs in construction engineering, engineering science, mining engineering, and petroleum engineering add to a lengthy list of career options in engineering alone. Coupled with thousands of programs in science and technical training offered at hundreds of other schools, the task of choosing the right field no doubt seems formidable.

Since you are reading this book, we assume that you are interested in studying engineering or at least are trying to decide whether or not to do so. Up to this point in your academic life, you have probably had little experience with engineering and have gathered your impressions of engineering from advertising materials, counselors, educators, and perhaps a practicing engineer or two. Now you must investigate as many careers as you can as soon as possible to be sure of making the right choice.

The study of engineering requires a strong background in mathematics and the physical sciences. Section 1.5 discusses typical areas of study within an engineering program that lead to the bachelor's degree. You should also consult with your counselor about specific course requirements. If you are enrolled in an engineering college but have not chosen a specific discipline, consult with an adviser or someone on the engineering faculty about particular course requirements in your areas of interest.

When considering a career in engineering or any closely related fields, you should explore the answers to several questions. What is engineering? What is an engineer? What are the func-

Figure 1.1
An engineering student
observes stress formation in an
automobile frame with the aid
of a virtual reality device. *(Iowa
State University.)*

tions of engineering? What are the engineering disciplines? Where does the engineer fit into the technical spectrum? How are engineers educated? What is meant by professionalism and engineering ethics? What have engineers done in the past? What are engineers doing now? What will engineers do in the future? Finding answers to such questions will assist you in assessing your educational goals and obtaining a clearer picture of the technological sphere.

Brief answers to some of these questions are given in this chapter. By no means are they intended to be a complete discussion of engineering and related fields. You can find additional and more detailed technical career information in the reference materials listed in the bibliography at the end of the book and by searching the World Wide Web.

1.2

The Technology Team

In 1876, 15 men led by Thomas Alva Edison gathered in Menlo Park, New Jersey, to work on "inventions." By 1887, the group had secured over 400 patents, including ones for the electric light bulb and the phonograph. Edison's approach typified that used for early engineering developments. Usually one person possessed nearly all the knowledge in one field and directed the research, development, design, and manufacture of new products in this field.

Today, however, technology has become so advanced and sophisticated that one person cannot possibly be aware of all the intricacies of a single device or process. The concept of systems engineering has thus evolved—that is, technological problems are studied and solved by a technology team.

Scientists, engineers, technologists, technicians, and craftspersons form the *technology team*. The functions of the team range across what is often called the *technical spectrum*. At one end of the spectrum are functions which involve work with scientific and engineering principles. At the other end of this technical spectrum are functions which bring designs into reality. Successful technology teams use the unique abilities of all team members to bring about a successful solution to a human need.

Each of the technology team members has a specific function in the technical spectrum, and it is of utmost importance that each specialist understand the role of all team members. It is not difficult to find instances where the education and tasks of team members overlap. For any engineering accomplishment, successful team performance requires cooperation that can be realized only through an understanding of the functions of the technology team. We will now investigate each of the team specialists in more detail. The technology team is one part of a larger team which has the overall responsibility for bringing a device, process, or system into reality. This team, frequently called a project or design team, may include, in addition to the technology team members, managers, sales representatives, field service persons, financial representatives, and purchasing personnel. These project teams meet frequently from the beginning of the project to insure that schedules and design specifications are met, and that potential problems are diagnosed early. This approach, intended to meet or exceed the customer's expectations, is referred to as total quality management (TQM) or continuous improvement (CI).

1.2.1
Scientist

Scientists have as their prime objective increased knowledge of nature (see Fig. 1.2). In the quest for new knowledge, the scientist conducts research in a systematic manner. The research steps referred to as the *scientific method* are often summarized as follows:

1. Formulate a hypothesis to explain a natural phenomenon.

2. Conceive and execute experiments to test the hypothesis.

3. Analyze test results and state conclusions.

4. Generalize the hypothesis into the form of a law or theory if experimental results are in harmony with the hypothesis.

5. Publish the new knowledge.

An open and inquisitive mind is an obvious characteristic of a scientist. Although the scientist's primary objective is that of obtaining an increased knowledge of nature, many scientists are also engaged in the development of their ideas into new and use-

Figure 1.2
Scientific research conducted in
an environmental laboratory.
(Iowa State University.)

ful creations. But to differentiate quite simply between the scientist and engineer, we might say that the true scientist seeks to understand more about natural phenomena, whereas the engineer primarily engages in applying new knowledge.

1.2.2
Engineer

The profession of engineering takes the knowledge of mathematics and natural sciences gained through study, experience, and practice and applies this knowledge with judgment to develop ways to utilize the materials and forces of nature for the benefit of all humans.

An engineer is a person who possesses this knowledge of mathematics and natural sciences, and through the principles of analysis and design, applies this knowledge to the solution of problems and the development of devices, processes, structures, and systems for the benefit of all humans.

Both the engineer and scientist are thoroughly educated in the mathematical and physical sciences, but the scientist primarily uses this knowledge to acquire new knowledge, whereas the engineer applies the knowledge to design and develop usable devices, structures, and processes. In other words, the scientist seeks to know, the engineer aims to do.

You might conclude that the engineer is totally dependent on the scientist for the knowledge to develop ideas for human benefit. Such is not always the case. Scientists learn a great deal from the work of engineers. For example, the science of thermodynamics was developed by a physicist from studies of practical steam engines built by engineers who had no science to guide them. On the other hand, engineers have applied the principles of nuclear fission discovered by scientists to develop nuclear power plants and numerous other devices and systems requiring nuclear reactions for their operation. The scientist's and engineer's functions frequently overlap, leading at times to a somewhat blurred image of the engineer. What distinguishes the engineer from the scientist in broad terms, however, is that the engineer often conducts research, but does so for the purpose of solving a problem.

The end result of an engineering effort—generally referred to as *design*—is a device, structure, system, or process which satisfies a need. A successful design is achieved when a logical procedure is followed to meet a specific need. The procedure, called the *design process,* is similar to the scientific method with respect to a step-by-step routine, but it differs in objectives and end results. The design process encompasses the following activities, all of which must be completed.

1. Identification of a need
2. Problem definition
3. Search
4. Constraints
5. Criteria
6. Alternative solutions
7. Analysis
8. Decision
9. Specification
10. Communication

In the majority of cases, designs are not accomplished by an engineer simply completing the 10 steps shown in the order given. As the designer proceeds through each step, new information may be discovered and new objectives may be specified for the design. If so, the designer must backtrack and repeat steps. For example, if none of the alternatives appear to be economically feasible when the final solution is to be selected, the designer must redefine the problem or possibly relax some of the criteria to admit less expensive alternatives. Thus, because decisions must frequently be made at each step as a result of new developments or unexpected outcomes, the design process becomes iterative.

It is very important that you begin your engineering studies with an appreciation of the thinking process used to arrive at a solution to a problem and ultimately to produce a successful result.

As you progress through your engineering education you will solve problems and learn the design process using the techniques of analysis and synthesis. Analysis is the act of separating a system into its constituent parts, whereas synthesis is the act of combining parts into a useful system. In the design process (Chapter 3), you will observe how analysis and synthesis are utilized to generate a solution to a human need.

Consider the cruise control in an automobile as a system. You can analyze the performance of this system by setting up a test under carefully controlled conditions—that is, you will define and control the operating environment for the system and note the performance of the system. For example, you may determine acceleration or deceleration when a speed change is requested by the driver. You may check to see if the speed returns to the desired level after braking to reduce the speed and using the resume control. During the design of a cruise control system, the system would be modeled on a computer, and performance would be predicted by adjusting the variables and observing the results through various graphical formats on the monitor. You can analyze the physical makeup of the cruise control by actually taking apart the control, identifying the parts according to form and function, and reassembling the control. In general, analysis is the taking of a system, establishing the operating environment, and determining the response (performance) of the system.

If you were attempting to design a new cruise control system, you would consider many methods for sensing speed, ways to adjust engine speed for acceleration and deceleration, ideas for driver interface with the control, and so forth. Many possible solutions will be generated, mostly in the form of conceptual solutions without the details. During the design phase, the computer model may be continually improved by "repeated analysis," that is, finding the best or optimum design by observing the effect of changes in the system variables. This is synthesis and is the inverse of analysis. Synthesis may be said to be the process of defining the desired response (performance) of a system, establishing the operating environment, and, from this, developing the system.

An example will illustrate.

Example problem 1.1 A protective liner exactly 12 m wide is available to line a channel for conveying water from a reservoir to downstream areas. If a trapezoidal-shaped channel (see Fig. 1.3) is constructed so that the liner will cover the surface completely, what is the flow area for $x = 2$ m and $\theta = 45°$? The geometry is defined such that $0 \leq x \leq 6$ and $0 \leq \theta \leq 90°$. Flow area multiplied by average flow velocity will yield volume rate of flow, an important parameter in the study of open-channel flows.

Solution The geometry is defined in Fig. 1.3. The flow area is given by the expression for the area of a trapezoid:

$$A = \tfrac{1}{2}(b_1 + b_2)h$$

where $b_1 = 12 - 2x$

$b_2 = 12 - 2x + 2x \cos \theta$

$h = x \sin \theta$

Figure 1.3

Therefore,

$$A = 12x \sin \theta - 2x^2 \sin \theta + x^2 \sin \theta \cos \theta$$

For the situation where $x = 2$ and $\theta = 45°$, the flow area is

$$A = (12)(2)(\sin 45°) - 2(2)^2(\sin 45°) + (2)^2(\sin 45°)(\cos 45°)$$

$$= 13.3 \text{ m}^2$$

Values of A can be quickly found for any combination of x and θ with a spreadsheet. Fig. 1.4a shows areas for $x = 2$ and a series of θ values. You have solved many problems of this nature by analysis; that is, a system is given (the channel as shown in Fig. 1.3), the operating environment is specified (the channel is flowing full), and you must find the system performance (determine the flow area). Analysis usually yields a unique solution.

Example problem 1.2 A protective liner exactly 12 m wide is available to line a channel conveying water from a reservoir to downstream areas. For the trapezoidal cross section shown in Fig. 1.3, what are the values of x and θ for a flow area of 16 m^2?

Solution Based on our work in Example prob. 1.1, we would have

$$16 = 12x \sin \theta - 2x^2 \sin \theta + x^2 \sin \theta \cos \theta$$

The solution procedure is not direct, and the solution is not unique, as it was in Example prob. 1.2. We begin our solution procedure by using a spreadsheet to generate a family of curves that illustrate the behavior of the implicit function of x and θ. Figure 1.4b shows the flow area as a function of θ for five values of x. We quickly observe that for $x = 1$ we cannot generate a flow area of 16 m^2. Also for $x = 4$ and 5, we definitely have two values of θ where a flow of 16 m^2 is possible. The spreadsheet will perform a search for the correct values. Figure 1.4c shows the result of a search between 0 and 90 at $x = 3$ m. In this situation a flow area of 16 m^2 occurs at a θ of 0.698014 radians or 40.0°. Other results are quickly obtainable by simply changing the value of x in the spreadsheet program in Fig. 1.4c.

You probably have not solved many problems of this nature. Example prob. 1.2 is a synthesis problem; that is, the operating environment is specified (channel flowing full), the performance is known (flow area is 16 m^2), and you must determine the system (values for x and θ). Example prob. 1.2 is the inverse problem to Example prob. 1.1. In general, synthesis problems do not have a unique solution, as can be seen from Example prob. 1.2.

Most of us have difficulty synthesizing. We cannot "see" a direct method to find an x and θ that yield a flow area of 16 m^2. Our solution to Example prob. 1.2 involved repeated analysis to "synthesize" the solution. We studied a family of curves (x is constant) of A versus θ, which enabled us to verify the spreadsheet analysis for values of x and θ that yield a specified flow area.

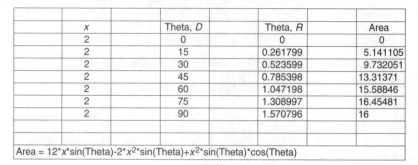

	x		Theta, D		Theta, R		Area
	2		0		0		0
	2		15		0.261799		5.141105
	2		30		0.523599		9.732051
	2		45		0.785398		13.31371
	2		60		1.047198		15.58846
	2		75		1.308997		16.45481
	2		90		1.570796		16

Area = 12*x*sin(Theta)-2*x^2*sin(Theta)+x^2*sin(Theta)*cos(Theta)

(a)

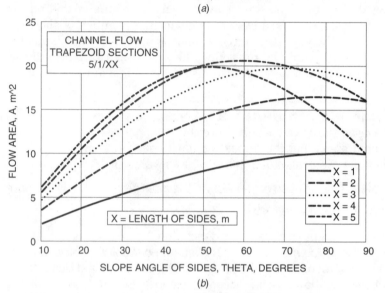

CHANNEL FLOW
TRAPEZOID SECTIONS
5/1/XX

X = LENGTH OF SIDES, m

FLOW AREA, A, m^2

SLOPE ANGLE OF SIDES, THETA, DEGREES

— X = 1
--- X = 2
··· X = 3
-·- X = 4
--- X = 5

(b)

	x		Theta,D		Theta, R		Area
	3		0		0		0
			90		1.570 796		18
	Area = 16 For Theta =				0.698014 Radians		

(c)

Theta, Degree		Theta, Radians		x, m			
0		0		1			
15		0.261799		2			
35		0.610865		3			
55		0.959931		4			
75		1.308997		5			
90		1.570796		6			
Max flow area		Theta		x			Area
		1.047198		4			20.78461

Area = 12*x*sin(Theta)-2*x^2*sin(Theta)+x^2*sin(Theta)*cos(Theta)

(d)

Figure 1.4

Example problem 1.3 For the situation described in Example prob. 1.2, find values of x and θ that yield a maximum flow area.

Solution   This is a design problem in which a "best" solution is sought, in this case a maximum flow area for the trapezoidal cross section shown in Fig. 1.3. We can determine the solution from the repeated analysis we did for Example prob. 1.2 The solution is obtained readily from a spreadsheet search over the range of values for x and θ. Figure 1.4d shows the result. You can verify this by checking against Fig. 1.4b.

Analysis and synthesis are very important to the engineering design effort, and a majority of your engineering education will involve techniques of analysis and synthesis in problem solving. We must not, however, forget the engineer's role in the entire design process. In an industrial setting, the objective is to correctly assess a need, determine the best solution to the need, and market the solution more quickly and less expensively than the competition. This demands careful adherence to the design process.

The successful engineer in a technology team will take advantage of computers and computer graphics. Today, with the aid of computers and computer graphics, it is possible to perform analysis, decide among alternatives, and communicate results far more quickly and with more accuracy than ever before. This translates into better engineering and an improved quality of living.

Terms like *computer-aided design* (CAD) and *computer-aided manufacturing* (CAM) label the modern engineering activities that continue to make engineering a challenging profession. Your work with analysis and synthesis techniques will require the use of a computer to a large extent in your education.

In addition to the use of the computer to perform computations and develop models, the information superhighway will provide you with instant access to new technologies, new processes, technical information, current economic conditions, and a multitude of other data that will help you achieve success in the workplace and in all other aspects of life. The Internet, a worldwide collection of computer networks, now has over 3 million servers which can be accessed for information. The World Wide Web (WWW) provides a user-friendly graphics interface to the Internet enabling text, audio, and video to be transmitted. Various search methods exist to help you find the information you are seeking.

Working engineers are now able to communicate with colleagues around the world via electronic mail (e-mail) on common interests and problems. We are able to monitor in real time a field test in a foreign nation of our design as we sit at our desks in the United States. With new company networks now being installed, called Intranets, databases of products, production status, design changes, and field status are at your fingertips.

In your personal life, you are able to join user groups with common interests in sports, music, home maintenance, automobile repair, and the like. The new paradigm of on-line, interactive control of the information we desire is unprecedented. We can get what information we want, when we want it. Not since the Gutenberg press has such a dramatic change occurred in the way we acquire and distribute knowledge.

One of your first tasks as an engineering student should be to locate a computer, find out how you can access the WWW, and learn how to navigate the Internet. You likely will be using this media to conduct research in your courses and to communicate with your instructors and classmates. If you are already capable of "surfing" the Internet through your own computer, you will find this to be most helpful in achieving your educational goals.

1.2.3
Technologist and Technician

Much of the actual work of converting the ideas of scientists and engineers into tangible results is performed by technologists and technicians (see Fig. 1.5). A technologist generally possesses a bachelor's degree and a technician an associate's degree. Technologists

Figure 1.5
Technologists work with engineers on the design and testing of a mobile soil characterization robot arm which, when equipped with a probe, gathers soil samples. *(Ames Laboratory, U.S. Department of Energy.)*

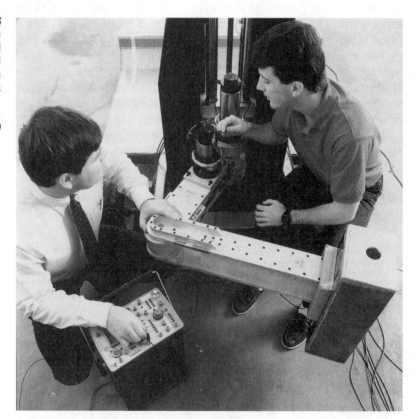

are involved in the direct application of their education and experience to make appropriate modifications in designs as the need arises. Technicians primarily perform computations and experiments and prepare design drawings as requested by engineers and scientists. Thus, technicians (typically) are educated in mathematics and science but not to the depth required of scientists and engineers. Technologists and technicians obtain a basic knowledge of engineering and scientific principles in a specific field and develop certain manual skills that enable them to communicate technically with all members of the technology team. Some tasks commonly performed by technologists and technicians include drafting, estimating, model building, data recording, and reduction, troubleshooting, servicing, and specification. Often they are the vital link between the idea on paper and the idea in practice.

1.2.4
Skilled Trades/Craftspersons

Members of the skilled trades possess the skills necessary to produce parts specified by scientists, engineers, technologists, and technicians. Craftspersons need not have an in-depth knowledge of the principles of science and engineering incorporated in a design (see Fig. 1.6). They are often trained on the job, serving an apprenticeship during which the skills and abilities to build and operate specialized equipment are developed. Some of the specialized jobs of craftspersons include those of welder, machinist, electrician, carpenter, plumber, and mason.

Figure 1.6
A machinist puts the finishing touches to a multiviewing transducer which will assist in locating the characterizing flaws in metals. *(Ames Laboratory, U.S. Department of Energy.)*

The Functions of the Engineer

As we alluded to in the previous section, engineering feats accomplished from earliest recorded history up to the industrial revolution could best be described as individual accomplishments. The various pyramids of Egypt were usually designed by one individual, who directed tens of thousands of laborers during construction. The person in charge called every move, made every decision, and took the credit if the project was successful or the consequences if the project failed.

With the industrial revolution, there was a rapid increase in scientific findings and technological advances. One-person engineering teams were no longer practical or desirable. We know that today no single aerospace engineer is responsible for the jumbo jets and no one civil engineer completely designs a bridge. Automobile manufacturers assign several thousand engineers to the design of a new model. So we not only have the technology team as described earlier, but we have engineers from many disciplines working together on single projects.

One approach to an explanation of an engineer's role in the technology spectrum is to describe the different types of work that engineers do. For example, civil, electrical, mechanical, and other engineers become involved in design, which is an engineering function. The *engineering functions,* which are discussed briefly in this section, are research, development, design, production, testing, construction, operations, sales, management, consulting, and teaching. Several of the *engineering disciplines* will be discussed in Sec. 1.4.

To avoid confusion between the meaning of the engineering disciplines and engineering functions, let us consider the following. Normally a student selects a curriculum (aerospace, chemical, mechanical, and so forth) either before or soon after admission to an engineering college. When and how the choice is made varies with each school. The point is, the student does not choose a function, but a discipline. To illustrate further, consider a student who has chosen mechanical engineering. This student will, during an undergraduate education, learn how mechanical engineers are involved in the engineering functions of research, development, design, and so on. Some program options allow a student to pursue an interest in a specific subdivision within the curriculum, such as energy conversion in a mechanical engineering program. Most other curricula have similar options.

Upon graduation, when you accept a job with a company, you will be assigned to a functional team performing in a specific area such as research, design, or sales. Within some companies, particularly smaller ones, you may become involved in more than one function—design *and* testing, for example. It is important to realize that regardless of your choice of discipline, you may become involved in one or more of the functions to be discussed in the following paragraphs.

1.3.1
Research

Successful research is one catalyst for starting the activities of a technology team or, in many cases, the activities of an entire industry. The research engineer seeks new findings, as does the scientist; but it must be kept in mind that the research engineer also seeks a way to use the discovery.

Key qualities of a successful research engineer are perceptiveness, patience, and self-confidence. Most students interested in research will pursue the master's and doctor's degrees in order to develop their intellectual abilities and the necessary research skills. An alert and perceptive mind is needed to recognize nature's truths when they are encountered. When attempting to reproduce natural phenomena in the laboratory, cleverness and patience are prime attributes. Research often involves tests, failures, retests, and so on, for long periods of time. Research engineers are therefore often discouraged and frustrated and must strain their abilities and rely on their self-confidence in order to sustain their efforts to a successful conclusion.

Billions of dollars are spent each year on research at colleges and universities, industrial research laboratories, government installations, and independent research institutes. The team approach to research is predominant today primarily because of the need to incorporate a vast amount of technical information into the research effort. Individual research is also carried out but not to the extent it was several years ago. A large share of research monies are channeled into the areas of energy, environment, health, defense, and space exploration. Research funding from fed-

Figure 1.7
Research engineers conducting an electronics experiment in an electrical engineering research laboratory. (*Iowa State University.*)

eral agencies is very sensitive to national and international priorities. During a career as a research engineer, you might expect to work in many diverse, seemingly unrelated areas, but your qualifications will allow you to adapt to many different research efforts.

1.3.2
Development

Using existing knowledge and new discoveries from research, the development engineer attempts to produce a device, structure, or process that is functional. Building and testing scale or pilot models is the primary means by which the development engineer evaluates ideas. A major portion of the development work requires use of well-known devices and processes in conjunction with established theories. Thus reading of available literature and a solid background in the sciences and in engineering principles are necessary for the development engineer's success.

Many people who suffer from heart irregularities are able to function normally today because of the pacemaker, an electronic device that maintains a regular heartbeat. The pacemaker is an excellent example of work of development engineers.

The first model, conceived by medical personnel and developed by engineers at the Electrodyne Company, was an external device that sent pulses of energy through electrodes to the heart. However, the power requirement for stimulus was so great that patients suffered severe burns on their chests. As improvements were being studied, research in surgery and electronics enabled development engineers to devise an external pacemaker with electrodes through the chest attached directly to the heart. Although more efficient from the standpoint of power require-

Figure 1.8
Development engineers use computer-aided design to assist the design and manufacture of prototype microprocessors.
(Iowa State University.)

ments, the devices were uncomfortable, and patients frequently suffered infection where the wires entered the chest. Finally, two independent teams developed the first internal pacemaker, 8 years after the original pacemaker had been tested. Their experience and research with tiny pulse generators for spacecraft led to this achievement. But the very fine wire used in these early models proved to be inadequate and quite often failed, forcing patients to have the entire pacemaker replaced. A team of engineers at General Electric developed a pacemaker that incorporated a new wire, called a *helicable*. The helicable consisted of 49 strands of wire coiled together and then wound into a spring. The spring diameter was about 46 μm (micrometers), half the diameter of a human hair. Thus, with doctors and development engineers working together, an effective, comfortable device was perfected that has enabled many heart patients to enjoy a more active life. Today pacemakers have been developed that operate at more than one speed, enabling the patient to speed up or slow down heart rate depending on physical activity.

We have discussed the pacemaker in detail to point out that changes in technology can be in part owing to development engineers. Only 13 years to develop an efficient, dependable pacemaker; 5 years to develop the transistor; 25 years to develop the digital computer—it only indicates that modern engineering methods generate and improve products nearly as fast as research generates new knowledge.

Successful development engineers are ingenious and creative. Astute judgment is often required in devising models that can be used to determine whether a project will be successful in performance and economical in production. Obtaining an advanced degree is helpful, but it is not as important as it is for an engineer who will be working in research. Practical experience more than anything else produces the qualities necessary for a career as a development engineer.

Development engineers are often asked to demonstrate that an idea will work. Within certain limits, they do not work out the exact specifications that a final product should possess. Such matters are usually left to the design engineer if the idea is deemed feasible.

1.3.3
Design

The development engineer produces a concept or model that is passed on to the design engineer for converting into a device, process, or structure (see Fig. 1.9). The designer relies on education and experience to evaluate many possible solutions, keeping in mind cost of manufacture, ease of production, availability of materials, and performance requirements. Usually several designs and redesigns will be undertaken before the product is brought before the general public.

Figure 1.9
Student design engineers verify
plans for a construction project.
(Iowa State University.)

To illustrate the role the design engineer plays, we will discuss the development of the over-the-shoulder seat belts for added safety in automobiles, which created something of a design problem. Designers had to decide where and how the anchors for the belts would be fastened to the car body. They had to determine what standard parts could be used and what parts had to be designed from scratch. Consideration was given to passenger comfort, inasmuch as awkward positioning could deter usage. Materials to be used for the anchors and the belt had to be selected. A retraction device had to be designed that would give flawless performance.

From one such part of a car, one can extrapolate the numerous considerations that must be given to the approximately 12,000 other parts that form the modern automobile: optimum placement of engine accessories, comfortable design of seats, maximization of trunk space, and aesthetically pleasing body design all require thousands of engineering hours to be successful in a highly competitive industry.

Like the development engineer, the designer is creative. However, unlike the development engineer, who is usually concerned only with a prototype or model, the designer is restricted by the state of the art in engineering materials, production facilities, and, perhaps most important, economic considerations. An excellent design from the standpoint of performance may be completely impractical when viewed from a monetary point of view. To make the necessary decisions, the designer must have a fundamental knowledge of many engineering specialty subjects as well as an understanding of economics and people.

1.3.4
Production and Testing

When research, development, and design have created a device for use by the public, the production and testing facilities are geared for mass production (see Figs. 1.10 and 1.11). The first

Figure 1.10
A test engineer observes the characteristics of an electronic circuit. (*Iowa State University.*)

step in production is to devise a schedule that will efficiently co-ordinate materials and personnel. The production engineer is re-sponsible for such tasks as ordering raw materials at the opti-mum times, setting up the assembly line, and handling and shipping the finished product. The individual who chooses this

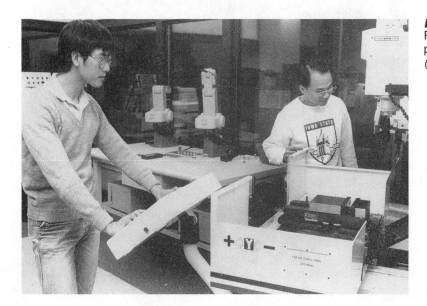

Figure 1.11
Production engineers design a prototype machining center. (*Iowa State University.*)

field must possess the ability to visualize the overall operation of a particular project as well as know each step of the production effort. Knowledge of design, economics, and psychology is of particular importance for production engineers.

Test engineers work with a product from the time it is conceived by the development engineer until such time as it may no longer be manufactured. In the automobile industry, for example, test engineers evaluate new devices and materials that may not appear in automobiles until several years from now. At the same time, they test component parts and completed cars currently coming off the assembly line. They are usually responsible for quality control of the manufacturing process. In addition to the education requirements of the design and production engineers, a fundamental knowledge of statistics is beneficial to the test engineer.

1.3.5
Construction

The counterpart of the production engineer in manufacturing is the construction engineer in the building industry (see Fig. 1.12). When an organization bids on a competitive construction project, the construction engineer begins the process by estimating material, labor, and overhead costs. If the bid is successful, a construction engineer assumes the responsibility of coordinating the project. On large projects, a team of construction engineers may supervise the individual segments of construction such as mechanical (plumbing), electrical (lighting), and civil (building). In

Figure 1.12
Construction engineers quite often work at the job site.
(Iowa State University.)

addition to a strong background in engineering fundamentals, the construction engineer needs on-the-job experience and an understanding of labor relations.

1.3.6
Operations

Up to this point, discussion has centered around the results of engineering efforts to discover, develop, design, and produce products that are of benefit to human beings. For such work, engineers obviously must have offices, laboratories, and production facilities in which to accomplish it. The major responsibility for supplying such facilities falls on the operations engineer (see Fig. 1.13). Sometimes called a plant engineer, this individual selects sites for facilities, specifies the layout for all facets of the operation, and selects the fixed equipment for climate control, lighting, and communication. Once the facility is in operation, the plant engineer is responsible for maintenance and modifications as requirements demand. Because this phase of engineering comes under the economic category of overhead, the operations

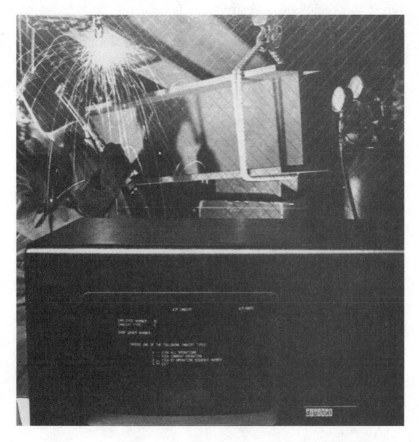

Figure 1.13
An operations engineer is responsible for putting together appropriate facilities such as this computer system used to direct manufacturing operations. *(Digital Equipment Corporation.)*

engineer must be very conscious of cost and keep up with new developments in equipment so that overhead is maintained at the lowest possible level. A knowledge of basic engineering, industrial engineering principles, economics, and law are prime educational requirements of the operations engineer.

1.3.7
Sales

In many respects, all engineers are involved in selling. To the research, development, design, production, construction, and operations engineer, selling means convincing management that money should be allocated for development of particular concepts or expansion of facilities. This is, in essence, selling one's own ideas. Sales engineering, however, means finding or creating a market for a product. The complexity of today's products requires an individual thoroughly familiar with materials in and operational procedures for consumer products to demonstrate to the consumer in layperson's terms how the products can be of benefit. The sales engineer is thus the liaison between the company and the consumer, a very important means of influencing a company's reputation. An engineering background plus a sincere interest in people and a desire to be helpful are the primary attributes of a sales engineer. The sales engineer usually spends a great deal of time in the plant learning about the product to be sold. After a customer purchases a product, the sales engineer is responsible for coordinating service and maintaining customer satisfaction. As important as sales engineering is to a company, it still has not received the interest from new engineering graduates that other engineering functions have. See Fig. 1.14.

Figure 1.14
A team of engineering design students making a sales presentation for their design.
(Iowa State University.)

Figure 1.15
A management engineer uses an electronic database to continually monitor the performance of the various units in a company. *(Iowa State University.)*

1.3.8
Management

Traditionally, management has consisted of individuals trained in business and groomed to assume positions leading to the top of the corporate ladder. However, with the influx of scientific and technological data being used in business plans and decisions, there has been a need for people in management with knowledge and experience in engineering and science. Recent trends indicate that a growing percentage of management positions are being assumed by engineers and scientists. Inasmuch as one of the principal functions of management is to use company facilities to produce an economically feasible product, and decisions must often be made that may affect thousands of people and involve millions of dollars over periods of several years, a balanced education of engineering or science and business seems to produce the best managerial potential.

1.3.9
Consulting

For someone interested in self-employment, a consulting position may be an attractive one (see Fig. 1.16). Consulting engineers operate alone or in partnership furnishing specialized help to clients who request it. Of course, as in any business, risks must be taken. Moreover, as in all engineering disciplines, a sense of integrity and a knack for correct engineering judgment are primary necessities in consulting.

A consulting engineer must possess a professional engineer's license before beginning practice. Consultants usually spend many years in a specific area before going on their own. A suc-

Figure 1.16
Consulting engineers designed the device being installed on a high-wire tower to measure the "gallop" in the wires caused by winds. *(Iowa State University.)*

cessful consulting engineer maintains a business primarily by being able to solve unique problems for which other companies have neither the time nor capacity. In many cases, large consulting firms maintain a staff of engineers of diverse backgrounds so that a wide range of engineering problems can be contracted.

1.3.10
Teaching

Individuals interested in a career that involves helping others to become engineers will find teaching very rewarding (see Fig. 1.17). The engineering teacher must possess an ability to communicate abstract principles and engineering experiences in a manner that young people can understand and appreciate. By merely following general guidelines, teachers are usually free to develop their own method of teaching and means of evaluating its effectiveness. In addition to teaching, the engineering educator can also become involved in student advising and research.

Engineering teachers today must have a mastery of fundamental engineering and science principles and a knowledge of applications. Customarily, they must obtain an advanced degree in order to improve their understanding of basic principles, to perform research in a specialized area, and perhaps to gain teaching experience on a part-time basis.

If you are interested in a teaching career in engineering or engineering technology, you should observe your teachers carefully as you pursue your degrees. Note how they approach the teaching process, the methodologies they use to stimulate learning, and their evaluation methods. Your initial teaching methods will likely be based on the best methods you observe as a student.

Figure 1.17
Teaching is a rewarding
profession with many
opportunities to help students
realize their educational goals.
(Iowa State University.)

1.4

The Engineering Disciplines

In the 1994 *Directory of Engineering and Engineering Technology* for undergraduate programs, 24 fields are listed for specialization. The opportunities to work in any of these areas or other specialty areas are numerous. Most engineering colleges offer some combination of the disciplines, primarily as 4-year programs. In some schools, two or more disciplines, such as industrial, management, and manufacturing engineering, are combined within one department which may offer separate degrees, or include one discipline as a specialty within another discipline. In this case a degree in the area of specialty is not offered. Other combinations of engineering disciplines include civil/construction/environmental and electrical/computer.

Figure 1.18 gives a breakdown of the number of engineering degrees in six categories for 1993–1994. Note that each category represents combined disciplines and does not provide information about a specific discipline within that category. The "other" category includes, among others, agricultural, biomedical, ceramic, metallurgical, mining, nuclear, safety, and ocean engineering.

Seven of the individual disciplines will be discussed in the following section. Engineering disciplines which pique your interest may be investigated in more detail by contacting the appropriate department or checking the library at your institution.

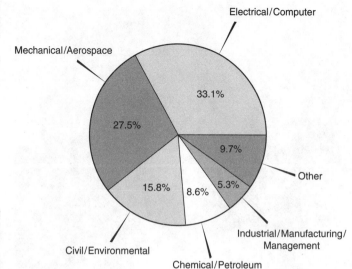

Electrical/Computer

Mechanical/Aerospace

33.1%

27.5%

9.7%

15.8%   8.6%   5.3%

Other

Figure 1.18
Engineering degrees by
discipline in 1993–94. Total
degrees awarded was 64,946.
(ASEE, PRISM, March 1995.)

Civil/Environmental

Chemical/Petroleum

Industrial/Manufacturing/
Management

1.4.1
Aerospace Engineering

Aerospace engineers study flight of all types of vehicles in all environments. They design, develop, and test aircraft, missiles, space vehicles, helicopters, hydrofoils, ships, and submerging ocean vehicles. The particular areas of specialty include aerodynamics, propulsion, orbital mechanics, stability and control, structures, design, and testing.

Aerodynamics is the study of the effects of moving a vehicle through the earth's atmosphere. The air produces forces that have both a positive effect on a properly designed vehicle (lift) and a negative effect (drag). In addition, at very high speeds the air generates heat on the vehicle which must be dissipated to protect crews, passengers, and cargo. Aerospace engineering students learn to determine such things as optimum wing and body shapes, vehicle performance, and environmental impact.

The operation and construction of turboprops, turbo and fan jets, rockets, ram and pulse jets, and nuclear and ion propulsion are part of the aerospace engineering student's study of propulsion. Such constraints as efficiency, noise levels, and flight distance enter into the selection of a propulsion system for a flight vehicle.

The aerospace engineer develops plans for interplanetary missions based on a knowledge of orbital mechanics. The problems encountered include determination of trajectories, stabilization, rendezvous with other vehicles, changes in orbit, and interception.

Stability and control involves the study of techniques for maintaining stability and establishing control of vehicles operating in the atmosphere or in space. Automatic control systems for autopilots and unmanned vehicles are part of the study of stability and control.

The study of structures is primarily involved with thin-shelled, flexible structures that can withstand high stresses and extreme temperature ranges. The structural engineer works closely with the aerodynamics engineer to determine the geometry of wings, fuselages, and control surfaces. The study of structures also involves thick-shelled structures that must withstand extreme pressures at ocean depths and lightweight composite structural materials for high performance vehicles.

The aerospace design engineer combines all the aspects of aerodynamics, propulsion, orbital mechanics, stability and control, and structures into the optimum vehicle. Design engineers work in a team and must learn to compromise in order to determine the best design satisfying all criteria and constraints.

The final proofing of a design involves the physical testing of a prototype. Aerospace test engineers learn to use testing devices such as wind tunnels, lasers, strain gauges, and data-acquisition systems. The testing takes place in structural laboratories, propulsion facilities, and in the flight medium with the actual vehicle.

1.4.2
Chemical Engineering

Chemical engineers deal with the chemical and physical principles that allow us to maintain a suitable environment. They create, design, and operate processes that produce useful materials including fuels, plastics, structural materials, food products, health products, fibers, and fertilizers. As our natural resources grow short, chemical engineers are creating substitutes or finding ways to extend our remaining resources.

The chemical engineer, in the development of new products, in designing processes, and in operating plants, may work in a laboratory, pilot plant, or full-scale plant. In the laboratory, the chemical engineer searches for new products and materials that

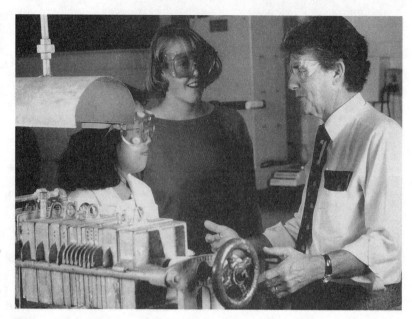

Figure 1.20
Process laboratories are a key
component in the education of
chemical engineers. *(Iowa State
University.)*

benefit humankind and the environment. This laboratory work
would be classified as research engineering.

In a pilot plant, the chemical engineer is trying to determine
the feasibility of carrying on a process on a large scale. There is
a great deal of difference between a process working in a test
tube in the laboratory and a process working in a production facility. The pilot plant is constructed to develop the necessary unit
operations to carry out the process. Unit operations are fundamental chemical and physical processes that are uniquely combined by the chemical engineer to produce the desired product.
A unit operation may involve separation of components by mechanical means such as filtering, settling, and floating. Separation
may also take place by changing the form of a component—for
example, through evaporation, absorption, or crystallization. Unit
operations also involve chemical reactions such as oxidation and
reduction. Certain chemical processes require the addition or re-
moval of heat or the transfer of mass. The chemical engineer thus
works with heat exchanges, furnaces, evaporators, condensers,
and refrigeration units in developing large-scale processes.

In a full-scale plant, the chemical engineer will continue to
"fine tune" the unit operations to produce the optimum process
based on the lowest cost. The day-by-day operations problems in
a chemical plant, such as piping, storage, and material handling,
are the responsibility of chemical engineers.

1.4.3
Civil Engineering

Civil engineering is the oldest branch of the engineering profession. The term *civil* was used to distinguish this field from mili-

tary engineers. Military engineers originated in Napoleon's army. The first engineers trained in this country were military engineers at West Point. Civil engineering involves application of the laws, forces, and materials of nature to the design, construction, operation, and maintenance of facilities that serve our needs in an efficient, economical manner. Civil engineers work for consulting firms engaged in private practice, for manufacturing firms, and for federal, state, and local governments. Because of the nature of their work, civil engineers assume a great deal of responsibility, which means that professional registration is an important goal for the civil engineer beginning practice. The specialties within civil engineering include structures, transportation, sanitary and water resources, geotechnical, surveying, and construction.

Structural engineers design bridges, buildings, dams, tunnels, and supporting structures. The designs include consideration of mass, winds, temperature extremes, and other natural phenomena such as earthquakes. Civil engineers with a strong structural background are often found in aerospace and manufacturing firms, playing an integral role in the design of vehicular structures.

Civil engineers in transportation plan, design, construct, operate, and maintain facilities that move people and goods throughout the world. For example, they make the decisions on where a freeway system should be located and describe the economic impact of the system on the affected public. They plan for growth of residential and industrial sectors of the nation. The modern rapid-transit systems are another example of the solution to a public need satisfied by transportation engineers.

Sanitary engineers are concerned with maintaining a healthful environment by proper treatment and distribution of water, treatment of wastewater, and control of all forms of pollution. The water resources engineer specializes in the evaluation of potential sources of new water for increasing or shifting populations, irrigation, and industrial needs.

Before any structure can be erected, a careful study of the soil, rock, and groundwater conditions must be undertaken to ensure stability. In addition to these studies, the geotechnical engineer analyzes building materials such as sand, gravel, and cement to determine proper consistency for concrete and other products.

Surveying engineers develop maps for any type of engineering project. For example, if a road is to be built through a mountain range, the surveyors will determine the exact route and develop the topographical survey which is then used by the transportation engineer to lay out the roadway.

Construction engineering is a significant portion of civil engineering, and many engineering colleges offer a separate degree in this area. Generally, construction engineers will work outside at the actual construction site. They become involved with the initial estimating of construction costs for surveying, excavation,

Figure 1.21
Civil engineering students
conducting a test on concrete
in the materials laboratory. *(Iowa
State University.)*

and construction. They will supervise the construction, start-up, and initial operation of the facility until the client is ready to assume operational responsibility. Construction engineers work around the world on many construction projects such as highways, skyscrapers, and power plants.

1.4.4
Electrical/Computer Engineering

Electrical/computer engineering is the largest branch of engineering, representing about 30 percent of the graduates entering the engineering profession. Because of the rapid advances in technology associated with electronics and computers, this branch of engineering is also the fastest growing.

The areas of specialty include communications, power, electronics, measurement, and computers.

We depend almost every minute of our lives on communication equipment developed by electrical engineers. Telephones, television, radio, and radar are common communications devices that we often take for granted. Our national defense system depends heavily on the communications engineer and on the hardware used for our early warning and detection systems.

The power engineer is responsible for producing and distributing the electricity demanded by residential, business, and industrial users throughout the world. The production of electricity requires a generating source such as fossil fuels, nuclear reactions, or hydroelectric dams. The power engineer may be involved with research and development of alternative generation sources such as sun, wind, and fuel cells. Transmission of electricity involves conductors and insulating materials. On the receiving end, appliances are designed by power engineers to be highly efficient in order to reduce both electrical demand and costs.

The area of electronics is the fastest-growing specialty in electrical engineering. The development of solid-state circuits (functional electronic circuits manufactured as one part rather than wired together) has produced high reliability in electronic devices. Microelectronics has revolutionized the computer industry and electronic controls. Circuit components much smaller than 1 micrometer (μm) in width enable reduced costs and higher electronic speeds to be attained in circuitry. The microprocessor, the principal component of a digital computer, is a major result of solid-state circuitry and microelectronics technology. The home computer, automobile control systems, and a multitude of electrical-applications devices conceived, designed, and produced by electronics engineers have greatly improved our standard of living.

Great strides have been made in the control and measurement of phenomena that occur in all types of processes. Physical quantities such as temperature, flow rate, stress, voltage, and acceleration are detected and displayed rapidly and accurately for optimal control of processes. In some cases, the data must be sensed at a remote location and accurately transmitted long distances to receiving stations. The determination of radiation levels is an example of the electrical process called *telemetry*.

The impact of microelectronics on the computer industry has created a multibillion dollar annual business that in turn has enhanced all other industries. The design, construction, and operation of computer systems is the task of computer engineers. This specialty within electrical engineering has in many schools become a separate degree program. Computer engineers deal with both hardware and software problems in the design and application of computer systems. The areas of application include re-

Figure 1.22
Computer engineering students enjoy a light moment with their professor in the software engineering laboratory. *(Iowa State University.)*

search, education, design engineering, scheduling, accounting, control of manufacturing operations, process control, and home computing needs. No single development in history has had as great an impact on our lives in as short a time span as has the computer.

1.4.5
Environmental Engineering

Environmental engineering deals with the appropriate use of our natural resources and the protection of our environment. For the most part, environmental engineering curricula are relatively new and in many instances reside as a specialty within other disciplines such as civil, chemical, and agricultural engineering. In the 1994 *Directory of Engineering and Engineering Technology* for undergraduate programs, 39 environmental engineering programs are listed as compared to 188 civil engineering programs. (Only ABET accredited programs are included in the directory.)

The construction, operation, and maintenance of the facilities we live and work in have a significant impact on the environment. Environmental engineers with a civil engineering background are instrumental in the design of water and wastewater treatment plants, facilities that resist natural disasters such as earthquakes and floods, and facilities that use no hazardous or toxic materials. The design and layout of large cities and urban areas must include protective measures for the disturbed environment.

Environmental engineers with a chemical engineering background are interested in air and water quality which is affected by many by-products of chemical and biological processes. Products which are slow to biodegrade are studied for recycling

Figure 1.23
Environmental engineers become involved in testing for contaminants at waste sites. *(Ames Laboratory, U.S. Department of Energy.)*

possibilities. Other products which may contaminate or be hazardous are being studied to develop better storage or to develop replacement products less dangerous to the environment.

With an agricultural engineering background, environmental engineers study air and water quality which is affected by animal production facilities, chemical run-off from agricultural fertilizers, and weed control chemicals. As we become more environmentally conscious, the demand for designs, processes, and structures which protect the environment will create an increasing demand for environmental engineers. They will provide the leadership for protecting our resources and environment for generations to come.

1.4.6
Industrial Engineering

Industrial engineering covers a broad spectrum of activities in organizations of all sizes. The principal efforts of industrial engineers are directed to the design of production systems for goods and services. As an example, consider the procedures and processes necessary to produce and market a power lawn mower. When the design of the lawn mower is complete, industrial engineers establish the manufacturing sequence from the point of bringing the materials to the manufacturing center to the final step of shipping the assembled lawn mowers to the marketing agencies. Industrial engineers develop a production schedule, oversee the ordering of standard parts (engines, wheels, bolts), develop a plant layout (assembly line) for production of nonstandard parts (frame, height adjustment mechanism), and perform a cost analysis for all phases of production.

As production is ongoing, industrial engineers will perform various studies, called *time and methods studies,* which assist in optimizing the handling of material, the shop processes, and the overall assembly line. In a large organization, industrial engineers will likely specialize in one of the many areas involved in the operation of a plant. In a smaller organization, industrial engineers are likely to be involved in all the plant activities. Because of their general study in many areas of engineering and their knowledge of the overall plant operations, industrial engineers are frequent choices for promotion into management-level positions.

The study of human factors is an important area of industrial engineering. In product design, for example, industrial engineers involved in fashioning automobile interiors study the comfort and fatigue factors of seats and instrumentation. And in the factory they develop training programs for operators and supervisors of new machinery or for new assembly-line operators.

With the rapid development of computer-aided manufacturing (CAM) techniques and of computer-integrated manufacturing

Figure 1.24
Industrial engineering students
program a pick-and-place robot
in an assembly process. *(Iowa
State University.)*

(CIM), the industrial engineer will play a large role in the factories
of the future. The industrial engineer of the future will also be involved in retraining the labor force to work in a high-technology
environment.

1.4.7
Mechanical Engineering

Mechanical engineering originated when people first began to use
levers, ropes, and pulleys to multiply their own strength and to
use wind and falling water as a source of energy. Today mechanical engineers are involved with all forms of energy utilization
and conversion, machines, manufacturing materials and processes,
and engines.

Mechanical engineers utilize energy in many ways for our benefit. Refrigeration systems keep perishable goods fresh for long
periods of time, air condition your homes and offices, and aid in
various forms of chemical processing. Heating and ventilating
systems keep us comfortable when the environment around us
changes with the seasons. Ventilating systems help keep the air
around us breathable by removing undesirable fumes. Mechanical
engineers analyze heat transfer from one object to another and
design heat exchangers to effect a desirable heat transfer.

The energy crisis of the 1970s brought to focus a need for new
sources of energy as well as new and improved methods of energy conversion. Mechanical engineers are involved in research in
solar, geothermal, and wind energy sources, along with research
to increase the efficiency of producing electricity from fossil fuel,
hydroelectric, and nuclear sources.

Machines and mechanisms used in all forms of manufacturing and transportation are designed and developed by mechanical engineers. Automobiles, airplanes, and trains combine a source of power and a machine to provide transportation. Tractors, combines, and other implements aid the agricultural community. Automated machinery and robotics are rapidly growing areas for mechanical engineers. Lathes, milling machines, grinders, and drills assist in the manufacture of goods. Sorting devices, typewriters, staplers, and mechanical pencils are part of the office environment. Machine design requires a strong mechanical engineering background and a vivid imagination.

In order to drive the machines, a source of power is needed. The mechanical engineer is involved with the generation of electricity by converting chemical energy in fuels to thermal energy in the form of steam, then to mechanical energy through a turbine to drive the electric generator. Internal-combustion devices such as gasoline, turbine, and diesel engines are designed for use in all areas of transportation. The mechanical engineer studies engine cycles, fuel requirements, ignition performance, power output, cooling systems, engine geometry, and lubrication in order to develop high-performance, low-energy-consuming engines.

The engines and machines designed by mechanical engineers require many types of materials for construction. The tools that are needed to process the raw material for other machines must be designed. For example, a very strong material is needed for a drill bit that must cut a hole in a steel plate. If the tool is made from steel it must be a higher-quality steel than that found in the plate. Methods of heat treating, tempering, and other metallurgical processes are applied by the mechanical engineer.

Figure 1.25
Mechanical engineering students test a robot in a manufacturing laboratory. *(Iowa State University.)*

Manufacturing processes such as electric-discharge machining, laser cutting, and modern welding methods are used by mechanical engineers in the development of improved products. Mechanical engineers are also involved in the testing of new materials and products such as composites.

1.5

Education of the Engineer

The amount of information coming from the academic and business world is increasing exponentially, and at the current rate it will double in less than 20 years. More than any other group, engineers are using this knowledge to shape civilization. To keep pace with a changing world, engineers must be educated to solve problems that are as yet unheard of. A large share of the responsibility for this mammoth education task falls on the engineering colleges and universities. But the completion of an engineering program is only the first step toward a lifetime of education. The engineer, with the assistance of the employer and the university, must continue to study. See Fig. 1.26.

Logically, then, an engineering education should provide a broad base in scientific and engineering principles, some study in the humanities and social sciences, and specialized studies in a chosen engineering curriculum. But specific questions concerning engineering education still arise. We will deal here with the questions that are frequently asked by students. What are the desirable characteristics for success in an engineering program? What knowledge and skills should be acquired in college? What is meant by continuing education with respect to an engineering career?

1.5.1
Desirable Characteristics

Years of experience have enabled engineering educators to analyze the performance of students in relation to abilities and de-

Figure 1.26
Long distance learning via satellite transmission makes it more convenient for working engineers to keep up on new developments in their field.
(Iowa State University.)

sires they possess entering college. The most important characteristics for an engineering student can be summarized as follows:

1. A strong interest in and ability to work with mathematics and science taken in high school.

2. An ability to think through a problem in a logical manner.

3. A knack for organizing and carrying through to conclusion the solution to a problem.

4. An unusual curiosity about how and why things work.

Although such attributes are desirable, having them is no guarantee of success in an engineering program. Simply a strong desire for the job has made successful engineers of some individuals who did not possess any of these characteristics; and, conversely, many who possessed them did not complete an engineering degree. Moreover, an engineering education is not easy, but it can offer a rewarding career to anyone who accepts the challenge.

1.5.2
Knowledge and Skills Required

As indicated previously, over 340 colleges and universities offer programs in engineering that are accredited by ABET or CEAB. These boards have as their purpose the quality control of engineering and technology programs offered in the United States and Canada. The basis of the boards is the engineering profession, which is represented through the participating professional groups. A listing of the participating bodies is given in Tab. 1.1.

The quality control of engineering programs is effected through the accreditation process. The engineering profession, through ABET and CEAB, has developed standards and criteria for the education of engineers entering the profession. Through visitations, evaluations, and reports, the written criteria and standards are compared with the engineering curricula at a university. Each program, if operating according to the standards and criteria, may receive up to 6 years of accreditation. If some discrepancies appear, accreditations may be granted for a shorter time period or may not be granted at all until appropriate improvements are made.

It is safe to say that for any given engineering discipline, no two schools will have identical offerings. However, close scrutiny will show a framework within which most courses can be placed, with differences occurring only in textbooks used, topics emphasized, and sequences followed. Figure 1.27 depicts this framework and some of the courses that fall within each of the areas. The approximate percentage of time spent on each course grouping is indicated.

The sociohumanistic block is a small portion of most engineering curricula, but it is important because it helps the engi-

Table 1.1 Participating bodies in the accreditation activity. (Compiled from ABET annual report, 1994.)

Organization	Abbreviation
American Academy of Environmental Engineers	AAEE
American Congress on Surveying and Mapping	ACSM
American Institute of Aeronautics and Astronautics, Inc.	AIAA
American Institute of Chemical Engineers	AICHE
American Nuclear Society	ANS
American Society of Agricultural Engineers	ASAE
American Society of Civil Engineers	ASCE
American Society for Engineering Education	ASEE
American Society of Heating, Refrigerating, and Air-Conditioning Engineers, Inc.	ASHRAE
The American Society of Mechanical Engineers	ASME
Institute of Industrial Engineers, Inc.	IIE
The Institute of Electrical and Electronics Engineers, Inc.	IEEE
The Minerals, Metals & Materials Society	TMS
National Council of Examiners for Engineering and Surveying	NCEES
National Institute of Ceramic Engineers	NICE
National Society of Professional Engineers	NSPE
Society of Automotive Engineers	SAE
Society of Manufacturing Engineers	SME
Society for Mining, Metallurgy, and Exploration, Inc.	SME-AIME
Society of Naval Architects and Marine Engineers	SNAME
Society of Petroleum Engineers	SPE

Figure 1.27
Elements of engineering curricula.

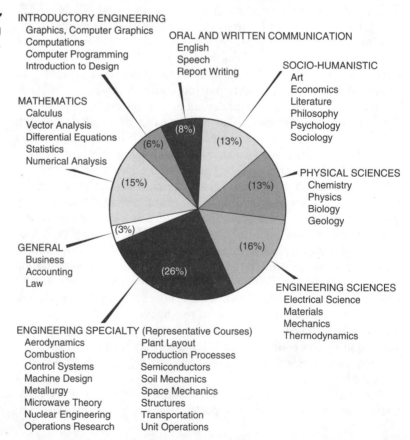

INTRODUCTORY ENGINEERING
Graphics, Computer Graphics
Computations
Computer Programming
Introduction to Design

ORAL AND WRITTEN COMMUNICATION
English
Speech
Report Writing

SOCIO-HUMANISTIC
Art
Economics
Literature
Philosophy
Psychology
Sociology

MATHEMATICS
Calculus
Vector Analysis
Differential Equations
Statistics
Numerical Analysis

PHYSICAL SCIENCES
Chemistry
Physics
Biology
Geology

GENERAL
Business
Accounting
Law

ENGINEERING SCIENCES
Electrical Science
Materials
Mechanics
Thermodynamics

ENGINEERING SPECIALTY (Representative Courses)
Aerodynamics   Plant Layout
Combustion   Production Processes
Control Systems   Semiconductors
Machine Design   Soil Mechanics
Metallurgy   Space Mechanics
Microwave Theory   Structures
Nuclear Engineering   Transportation
Operations Research   Unit Operations

(8%) (13%) (6%) (13%) (15%) (3%) (16%) (26%)

neering student understand and develop an appreciation for the potential impact of engineering undertakings on the environment and general society. When the location of a nuclear power plant is being considered, the engineers involved in this decision must respect the concerns and feelings of all individuals who might be affected by the location. Discussions of the interaction between engineers and the general public take place in few engineering courses; sociohumanistic courses are thus needed to furnish engineering students with an insight into the needs and aspirations of society.

Chemistry and physics are almost universally required in engineering. They are fundamental to the study of engineering science. The mathematics normally required for college chemistry and physics is more advanced than that for the corresponding high school courses. Higher-level chemistry and physics may also be required, depending on departmental structure. Finally, other physical science courses may be required in some programs or may be taken as electives.

An engineer cannot be successful without the ability to communicate ideas and the results of work efforts. The research engineer writes reports and orally presents ideas to management. The production engineer must be able to converse with craftspersons in understandable terms. And all engineers have dealings with the public and must be able to communicate on a nontechnical level.

Engineers have been accused of not becoming involved in public affairs. The reason often given for their not becoming involved is that they are not trained sufficiently in oral and written communications. However, the equivalent of one-third of 1 year is spent on formal courses in these these subjects, and additional time is spent in design presentations, written laboratory reports, and the like. A conscious effort by student engineers must be made to improve their abilities in oral and written communication to overcome this nonactivist label.

Mathematics is the most powerful tool the engineer uses to solve problems. The amount of time spent in this area is indicative of its importance. Courses in calculus, vector analysis, and differential equations are common to all degree programs. Statistics, numerical analysis, and other mathematics courses support some engineering specialty areas. Students desiring an advanced degree may want to take mathematics courses beyond the baccalaureate-level requirements.

In the early stages of an engineering education, introductory courses in graphical communication, computational techniques, design, and computer programming are taken. Engineering schools vary somewhat in their emphasis on these areas, but the general intent is to develop skills in the application of theory to practical problem solving and familiarity with engineering terminology. Design is presented from a conceptual point of view to aid the student in creative thinking. Graphics develops the visualization capability and assists the student in transferring mental thoughts into well-defined concepts on paper. The tremendous potential of the computer to assist the engineer has led to the requirement of computer programming in almost all curricula. Use of the computer to perform many tedious calculations has increased the efficiency of the engineer and allows more time for creative thinking. Computer graphics is becoming a part of engineering curricula. Its ability to enhance the visualization of geometry and to depict engineering quantities graphically has increased productivity in the design process.

With a strong background in mathematics and physical sciences, you can begin study of engineering sciences, courses that are fundamental to all engineering specialties. Electrical science includes study of charges, fields, circuits, and electronics. Materials science courses involve study of the properties and chemical compositions of metallic and nonmetallic substances (see Fig. 1.28). Mechanics includes study of statics, dynamics, fluids, and mechanics of materials. Thermodynamics is the science of heat and is the basis for the study of all types of energy and energy transfer. A sound understanding of the engineering sciences is most important for anyone interested in pursuing postgraduate work and research.

Figure 1.27 shows only a few examples of the many specialized engineering courses given. Scanning course descriptions in a college general bulletin or catalog will provide a more detailed insight into the specialized courses required in the various engineering disciplines.

Most curricula allow a student flexibility in selecting a few courses in areas not previously mentioned. For example, a stu-

Figure 1.28
Laboratories are instrumental in the learning process for student engineers. *(Iowa State University.)*

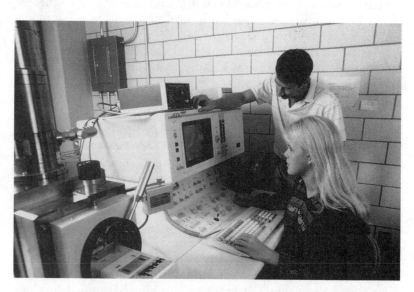

dent interested in management may take some courses in business and accounting. Another may desire some background in law or medicine, with the intent of entering a professional school in one of these areas upon graduation from engineering.

Engineering is a learned vocation, demanding an individual with high standards of ethics and sound moral character. When making judgments that may create controversy and affect many people, the engineer must keep foremost in mind a dedication to the betterment of humanity.

The engineering profession has attempted for many years to become unified. However, technical societies that represent individual engineering disciplines have grown strong and tend to keep the various engineering disciplines separated. This contrasts with more unified professions such as law, medicine, and theology.

1.6.1 Professionalism

Professionalism is a way of life. A professional person is one who engages in an activity that requires a specialized and comprehensive education and is motivated by a strong desire to serve humanity. A professional thinks and acts in a manner that brings favor upon the individual and the entire profession. Developing a professional frame of mind begins with your engineering education.

The professional engineer can be said to have the following:

1. Specialized knowledge and skills used for the benefit of humanity.

2. Honesty and impartiality in engineering service.

3. Constant interest in improving the profession.

4. Support of professional and technical societies that represent the professional engineer.

It is clear that these characteristics include not only technical competence but also a positive attitude toward life that is continually reinforced by educational accomplishments and professional service.

A primary reason for the rapid development in science and engineering is the work of the technical societies. The fundamental service provided by a society is the sharing of ideas, which means that technical specialists can publicize their efforts and assist others in promoting excellence in the profession. When information is distributed to other society members, new ideas evolve and duplicated efforts are minimized. The societies conduct meetings on international, national, and local bases. Students of engineering will find a technical society in their specialty that may operate as a branch of the regular society or as a student chapter on campus. The student organization is an important link with the professional workers, providing motivation and the opportunity to make acquaintances that will help students to formulate career objectives.

Table 1.1 is a partial listing of the numerous engineering societies that support the engineering disciplines and functions. These technical societies are linked because of their support of the accreditation process. Over 60 other societies exist for the purpose of supporting the professional status of engineers. Among these are the Society of Women Engineers (SWE), the National Society of Black Engineers (NSBE), the Acoustical Society of America (ASA), the Society of Plastics Engineers (SPE), and the American Society for Quality Control (ASQC).

Many of the technical societies are quite influential in the engineering profession. To unify the profession in the manner of such other professions as law, medicine, and theology would require the cooperation of all the societies. However the individual societies have more than satisfied the professional needs for many engineers, so no pressing desire to unify is apparent. Nonetheless, to preserve the advantages of the technical societies while unifying the entire profession remains a long-range goal of engineers today.

1.6.2
Professional Registration

The power to license engineers rests with each of the 50 states. Since the first registration law in Wyoming in 1907, all states have developed legislation specifying requirements for engineering practice. The purpose of registration laws is to protect the public. Just as one would expect a physician to provide competent medical service, an engineer can be expected to provide competent technical service. However, the laws of registration for engineers are quite different from those for lawyers or physicians. An engineer does not have to be registered to practice engineering. Legally, only the chief engineer of a firm needs to be registered for that firm to perform engineering services. Individuals testifying as expert engineering witnesses in court and individuals offering engineering consulting services need to be registered. In some instances, the practice of engineering is allowed as long as the individual does not advertise as an engineer.

The legal process for becoming a licensed professional engineer consists of four parts, two of which entail examinations. The parts include:

1. An engineering degree from an acceptable institution as defined by the state board for registration. Graduation from an ABET-accredited institution satisfies the degree requirement automatically.

2. Successful completion of the Fundamentals of Engineering Examination (FE) entitles one to the title "engineer-in-training" (EIT). This 8-hour exam may be taken during the last term of an undergraduate program that is ABET-accredited. The first half of the exam covers fundamentals in the areas of mathematics, chemistry, physics, engineering mechanics, electrical science, thermal science, economics, and ethics. The second half is oriented to each discipline, such as mechanical. The passing grade is determined by the state board.

3. Completion of 4 years of engineering practice as an EIT.

4. Successful completion of the Principles and Practice Examination completes the licensing process. This is also an 8-hour examination covering problems normally encountered in the area of specialty, such as mechanical or chemical engineering.

It should be noted that once the license is received, it is permanent, although there is an annual renewal fee. In addition, there is a trend toward specific requirements in continuing education each year in order to maintain the license. Licensed engineers in some states may attend professional meetings in their specialty, take classes, and write professional papers or books to accumulate sufficient professional development activities beyond their job responsibilities to maintain their licenses. This trend is a reflection of the rapidly changing technology and the need for engineers to remain current in their area.

Registration does have many advantages. Most public employment positions, all expert-witness roles in court cases, and some high-level company positions require the professional engineer's license. However, less than one-half of the eligible candidates are currently registered. You should give serious consideration to becoming registered as soon as you qualify. Satisfying the requirements for registration can be started even before graduation from an ABET-accredited curriculum.

1.6.3
Professional Ethics

Ethics is the guide to personal conduct of a professional. Most of the technical societies have a written code of ethics for their members. Because of this, some variations exist; but a general view of ethics for engineers is provided here for two of the technical societies. Figure 1.29 is a code endorsed by the Accreditation Board for Engineering and Technology. Figure 1.30 is the "Engineer's Creed" as published by the National Society of Professional Engineers.

CODE OF ETHICS OF ENGINEERS

THE FUNDAMENTAL PRINCIPLES

Engineers uphold and advance the integrity, honor and dignity of the engineering profession by:

I. using their knowledge and skill for the enhancement of human welfare;

II. being honest and impartial, and serving with fidelity the public, their employers and clients;

III. striving to increase the competence and prestige of the engineering profession; and

IV. supporting the professional and technical societies of their disciplines.

THE FUNDAMENTAL CANONS

1. Engineers shall hold paramount the safety, health and welfare of the public in the performance of their professional duties.

2. Engineers shall perform services only in the areas of their competence.

3. Engineers shall issue public statements only in an objective and truthful manner.

3. Engineers shall act in professional matters for each employer or client as faithful agents or trustees, and shall avoid conflicts of interest.

5. Engineers shall build their professional reputation on the merit of their services and shall not compete unfairly with others.

6. Engineers shall act in such a manner as to uphold and enhance the honor, integrity and dignity of the profession.

7. Engineers shall continue their professional development throughout their careers and shall provide opportunities for the professional development of those engineers under their supervision.

345 East 47th Street New York, NY 10017

*Formerly Engineers' Council for Professional Development. (Approved by the ECPD Board of Directors, October 5, 1977.)

Figure 1.29
Code of Ethics for Engineers. *(Accreditation Board for Engineering and Technology.)*

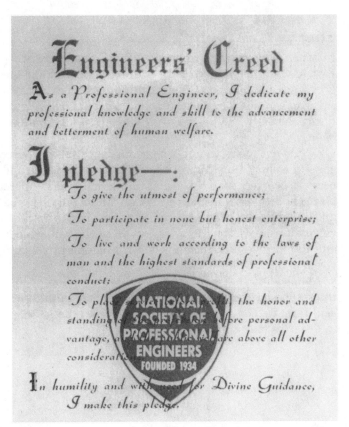

1.7

Challenges of the Future

The world continues to undergo rapid and sometimes tumultuous change. As a practicing engineer, you will occupy center stage in many of these changes in the near future and will become even more involved in the more distant future. The huge tasks of providing energy, maintaining a supply of water, ensuring a competitive edge in the world marketplace, rebuilding our infrastructure, and preserving our environment will challenge the technical community beyond anyone's imagination.

Engineers of today have nearly instantaneous access to a wealth of information from technical, economic, social, and political sources. A key to the success of engineers in the future will be the ability to study and absorb the appropriate information in the time allotted for producing a design or solution to a problem. A degree in engineering is only the beginning of a lifelong period of study in order to remain informed and competent in the field.

Engineers of tomorrow will have even greater access to information and will use increasingly powerful computer systems to digest this information. They will work with colleagues around the world solving problems and creating new products. They will assume greater roles in making decisions that affect the use of energy, water, and other natural resources.

1.7.1
Energy

In order to develop technologically, nations of the world require vast amounts of energy. With a finite supply of our greatest energy source, fossil fuels, alternate supplies must be developed and existing sources must be controlled with a worldwide usage plan. A key factor in the design of products must be minimum use of energy.

We do not mean to imply that our fossil fuel resources will be gone in a short time. However, as demand increases and supplies become scarcer, the cost of obtaining the energy increases and places additional burdens on already financially strapped regions and individuals. Engineers with great vision are needed to develop alternative sources of energy from the sun, wind, and ocean and to improve the efficiency of existing energy consumption devices.

Along with the production and consumption of energy come the secondary problems of pollution. Such pollutants as smog and acid rain, carbon monoxide, and radiation must receive attention in order to maintain the balance of nature.

Figure 1.31

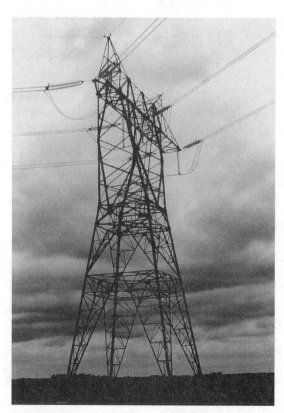

1.7.2
Water

The basic water cycle—from evaporation to cloud formation, then
to rain, runoff, and evaporation again—is taken for granted by
most people. However, if the rain or the runoff is polluted, then
the cycle is interrupted and our water supply becomes a crucial
problem. In addition, some highly populated areas have a limited
water supply and must rely on water distribution systems from
other areas of the country. Many formerly undeveloped agricultural regions are now productive because of irrigation systems.
However, the irrigation systems deplete the underground streams
of water that are needed downstream.

These problems must be solved in order for life to continue to
exist as we know it. Because of the regional water distribution
patterns, the federal government must be a part of the decision-
making process for water distribution. One of the concerns that
must be eased is the amount of time required to bring a water
distribution plan into effect. Government agencies and the private sector are strapped by regulations that cause delays in planning and construction of several years. Greater cooperation and
a better informed public are goals that public works engineers
must strive to achieve.

Developing nations around the world need an additional water
supply because of increasing population. Many of these nations
do not have the necessary fresh water and must rely on desali-
nation, a costly process. The continued need for water is a concern for leaders of the world, and engineers will be asked to
create additional sources of this life-sustaining resource.

1.7.3
A Competitive Edge in the World
Marketplace

We have all purchased or used products that were manufactured
outside the country. Many of these products incorporate technology that was developed in the United States. In order to maintain our strong industrial base, we must develop practices and
processes that enable us to compete, not just with other U.S. industries, but with international industries.

The goal of any industry is to generate a profit. In today's marketplace this means creating the best product in the shortest time
at a lesser price than the competition. A modern design process
incorporating sophisticated analysis procedures and supported by
high-speed computers with graphical displays increases the capability for developing the "best" product. The concept of inte-
grating the design and manufacturing functions with CAD/CAM
and CIM promises to shorten the design-to-market time for new
products and for upgraded versions of existing products. The de-

velopment of the automated factory is an exciting concept that is receiving a great deal of attention from manufacturing engineers today. Remaining competitive by producing at a lesser price requires a national effort involving labor, government, and distribution factors. In any case, engineers are going to have a significant role in the future of our industrial sector.

1.7.4
Infrastructure

All societies depend upon an infrastructure of transportation, waste disposal, and water distribution systems for the benefit of the population. In the United States much of infrastructure is in a state of deterioration without sound plans for upgrading. For example,

1. Commercial jet fleets include aircraft that are 25 to 30 years old. Major programs are now underway to safely extend the service life of these jets. In order to survive economically, airlines must balance new replacement jets with a program to keep older planes flying.

2. One-half of the sewage treatment plants cannot satisfactorily handle the demand.

3. The interstate highway system, nearing 40 years old in many areas, needs major repairs throughout. Local paved roads are deteriorating because of a lack of infrastructure funds.

4. Many bridges are potentially dangerous to the traffic loads on them.

5. Railroads continue to struggle with maintenance of the railbeds and the rolling stock in the face of stiff commercial competition from the air freight and truck transportation industries.

6. Municipal water systems require billions of dollars in repairs and upgrades to meet the demands of the public and the stiffening water quality requirements.

Figure 1.32
As our infrastructure ages, we must earmark the funds necessary to upgrade, expand, and repair these facilities to provide for our needs.

It is estimated that the total value of the public works facilities is over $2 trillion. To protect this investment, innovative thinking and creative funding must be fostered. Some of this is already occurring in road design and repair. For example, a method has been applied successfully that recycles asphalt pavement and actually produces a stronger product. Engineering research is producing extended-life pavement with new additives and structural designs. New, relatively inexpensive methods of strengthening old bridges have been used successfully.

1.7.5
Environment

Our insatiable demand for energy, water, and other national resources creates imbalances in nature which only time and serious conservation efforts can keep under control. The concern for environmental quality is focused in four areas: cleanup, compliance, conservation, and pollution prevention. Partnerships between industry, government, and consumers are working to establish guidelines and regulations in the gathering of raw materials, manufacturing of consumer products, and the disposal of material at the end of its designed use.

The American Plastics Council publishes a document entitled *Designing for the Environment* which describes environmental issues and initiatives affecting product design. All engineers need to be aware of these initiatives and how they apply in their particular industries:

- Design for the environment (DFE)—incorporate environmental considerations into product designs to promote environmental stewardship.
- Environmentally conscious manufacturing (ECM) or green manufacturing—incorporate pollution prevention and toxics use reduction in the making of products to promote environmental stewardship.
- Extended product or producer responsibility—product manufacturers assume responsibility for taking back their products at the end of the products' life and disposing of them according to defined environmental criteria.
- Life-cycle assessment (LCA)—quantified assessment of the environmental impacts associated with all phases of a product's life, often from the extraction of base minerals through the life of the product.
- Pollution prevention—reduce pollution sources in the design phase of products instead of addressing pollution after it is generated.
- Toxic use reduction—reduce the amount, toxicity, and number of toxic chemicals used in manufacturing.

As you can see from these initiatives all engineers, regardless of discipline, must be environmentally conscious in their work. In the next few decades, we will face tough decisions regarding our environment. Engineers will play a major role in making the correct decisions for our small, delicate world.

1.7.6
Conclusion

We have touched only briefly on the possibilities for exciting and rewarding work in all engineering areas. The first step is to obtain the knowledge during your college education that is necessary for your first technical position. After that, you must continue your education, either formally by seeking an advanced degree or degrees, or informally through continuing education courses or appropriate reading to maintain pace with the technology, an absolute necessity for a professional. Many challenges await you. Prepare to meet them well.

Problems

1.1 Compare the definitions of an engineer and a scientist from at least three different sources.

1.2 Compare the definitions of an engineer and a technologist from at least two sources.

1.3 Find three textbooks that introduce the design process. Copy the steps in the process from each textbook. Note similarities and differences and write a paragraph describing your conclusions.

1.4 Find the name of a pioneer engineer in the field of your choice and write a brief paper on the accomplishments of this individual.

1.5 Select a specific branch of engineering and list at least 20 different industrial organizations that utilize engineers from this field.

1.6 Select a branch of engineering, such as mechanical engineering, and an engineering function, such as design. Write a brief paper on some typical activities that are undertaken by the engineer performing the specified function. Sources of information can include books on engineering career opportunities and practicing engineers in the particular branch.

1.7 For a particular branch of engineering, such as electrical engineering, find the program of study for the first 2 years and compare it with the program offered at your school approximately 20 years ago. Comment on the major differences.

1.8 Do Prob. 1.7 for the *last* 2 years of study in a particular branch of engineering.

1.9 List five of your own personal characteristics and compare that list with the list in Sec. 1.5.1.

1.10 Prepare a brief paper on the requirements for professional registration in your state. Include the type and content of the required examinations.

1.11 Prepare a 5-minute talk to present to your class describing one of the technical societies and how it can benefit you as a student.

1.12 Choose one of the following topics (or one suggested by your instructor) and write a paper that discusses technological changes that have occurred in this area in the past 15 years. Include commentary on the social and environmental impact of the changes and on new problems that may have arisen because of the changes.

    (*a*) passenger automobiles
    (*b*) electric power-generating plants
    (*c*) computer graphics
    (*d*) heart surgery
    (*e*) heating systems (furnaces)
    (*f*) microprocessors
    (*g*) water treatment
    (*h*) road paving (both concrete and asphalt)
    (*i*) computer-controlled metal fabrication processes
    (*j*) robotics
    (*k*) air conditioning

Engineering Solutions

The practice of engineering involves the application of accumulated knowledge and experience to a wide variety of technical situations. Two areas, in particular, that are fundamental to all of engineering are design and problem solving. The professional engineer is expected to intelligently and efficiently approach, analyze, and solve a range of technical problems. These problems can vary from single solution, reasonably simple problems to extremely complex, open-ended problems that require a multidisciplinary team of engineers.

Problem solving is a combination of experience, knowledge, process, and art. Most engineers either through training or experience solve many problem by process. The design process, for example, is a series of logical steps that when followed produce an optimal solution given time and resources as two constraints. The total quality (TQ) method is another example of a process. This concept suggests a series of steps leading to desired results while exceeding customer expectations.

This chapter provides a basic guide to problem analysis, organization, and presentation. Early in your education, you must develop an ability to solve and present simple or complex problems in an orderly, logical, and systematic way.

A distinguishing characteristic of a qualified engineer is the ability to solve problems. Mastery of problem solving involves a combination of art and science. By *science* we mean a knowledge of the principles of mathematics, chemistry, physics, mechanics, and other technical subjects that must be learned so that they can be applied correctly. By *art* we mean the proper judgment, experience, common sense, and know-how that must be used to reduce a real-life problem to such a form that science can be applied to its solution. To know when and how rigorously science should be applied and whether the resulting answer reasonably satisfies the original problem is an art.

Much of the science of successful problem solving comes from formal education in school or from continuing education after graduation. But most of the art of problem solving cannot be learned in a formal course; rather, it is a result of experience and common sense. Its application can be more effective, however, if problem solving is approached in a logical and organized method—that is, if it follows a process.

To clarify the distinction, let us suppose that a manufacturing engineer working for an electronics company is given the task of recommending whether the introduction of a new personal computer that will focus on an inexpensive system for the home market can be profitably produced. At the time the engineering task is assigned, the competitive selling price has already been established by the marketing division. Also, the design group has developed working models of the personal computer with specifications of all components, which means that the cost of these components is known. The question of profit thus rests on the cost of assembly. The theory of engineering economy (the science portion of problem solving) is well known by the engineer and is applicable to the cost factors and time frame involved. Once the production methods have been established, the cost of assembly can be computed using standard techniques. Selection of production methods (the art portion of problem solving) depends largely on the experience of the engineer. Knowing what will or will not work in each part of the manufacturing process is a must in the cost estimate, but that data cannot be found in a handbook. It is in the mind of the engineer. It is an art originating from experience, common sense, and good judgment.

Before the solution to any problem is undertaken, whether by a student or by a practicing professional engineer, a number of important ideas must be considered. Consider the following questions: How important is the answer to a given problem? Would a rough, preliminary estimate be satisfactory, or is a high degree of accuracy demanded? How much time do you have and what resources are at your disposal? In an actual situation your answers may depend on the amount of data available or the amount that must be collected, the sophistication of equipment that must be used, the accuracy of the data, the number of people available to assist, and many other factors. Most complex problems require some level of computer support. What about the theory you intend to use? Is it state of the art? Is it valid for this particular application? Do you currently understand the theory, or must time be allocated for review and learning? Can you make assumptions that simplify without sacrificing needed accuracy? Are other assumptions valid and applicable?

The art of problem solving is a skill developed with practice. It is the ability to arrive at a proper balance between the time and resources expended on a problem and the accuracy and validity obtained in the solution. When you can optimize time and resources versus reliability, then problem-solving skills will serve you well.

The engineering method is an example of process. Earlier the engineering design process was mentioned. Although there are different processes that could be listed, a typical process is represented by the following 10 steps:

1. Identify the problem
2. Define the problem
3. Search
4. Constraints
5. Criteria
6. Alternative solutions
7. Analysis
8. Decision
9. Specification
10. Communication

These design steps are simply the overall process that an engineer uses when solving an open-ended problem. One significant portion of this design procedure is step 7—analysis.

Analysis is the use of mathematical and scientific principles to verify the performance of alternative solutions. Analyses conducted by engineers in many design projects normally involve three areas: application of the laws of nature, application of the laws of economics, and application of common sense.

The analysis phase can be used as an example to demonstrate how cyclic the design process is intended to be. Within any given step or phase there are still other processes that can be applied. One very important such process is called the *engineering method*. It consists of six basic steps:

1. Recognize and Understand the Problem

Perhaps the most difficult part of problem solving is developing the ability to recognize and define the problem precisely. This is true at the beginning of the design process and when applying the engineering method to a subpart of the overall problem. Many academic problems that you will be asked to solve have this step completed by the instructor. For example, if your instructor asks you to solve a quadratic algebraic equation but provides you with all the coefficients, the problem has been completely defined before it is given to you and little doubt remains about what the problem is.

If the problem is not well defined, considerable effort must be expended at the beginning in studying the problem, eliminating the things that are unimportant, and focusing on the root problem. Effort at this step pays great dividends by eliminating or reducing false trials, thereby shortening the time taken to complete later steps.

2. Accumulate Data and Verify Accuracy

All pertinent physical facts such as sizes, temperatures, voltages, currents, costs, concentrations, weights, times, and so on must be ascertained. Some problems require that steps 1 and 2 be done simultaneously. In others, step 1 might automatically produce some of the physical facts. Do not mix or confuse these details with data that are suspect or only assumed to be accurate. Deal only with items that can be verified. Sometimes it will pay to actually verify data that you believe to be factual but may actually be in error.

3. Select the Appropriate Theory or Principle

Select appropriate theories or scientific principles that apply to the solution of the problem; understand and identify limitations or constraints that apply to the selected theory.

4. Make Necessary Assumptions

Perfect solutions do not exist to real problems. Simplifications need to be made if they are to be solved. Certain assumptions can be made that do not significantly affect the accuracy of the solution, yet other assumptions may result in a large reduction in accuracy.

Although the selection of a theory or principle is stated in the engineering method as preceding the introduction of simplifying assumptions, there are cases where the order of these two steps should be reversed. For example, if you were solving a material balance problem you often need to assume that the process is steady, uniform, and without chemical reactions, so that the applicable theory can be simplified.

5. Solve the Problem

If steps 3 and 4 have resulted in a mathematical equation (model), it is normally solved by application of mathematical theory, although a trial-and-error solution which employs the use of a computer or perhaps some form of graphical solution may also be applicable. The results will normally be in numerical form with appropriate units.

6. Verify and Check Results

In engineering practice, the work is not finished merely because a solution has been obtained. It must be checked to ensure that it is mathematically correct and that units have been properly specified. Correctness can be verified by reworking the problem using a different technique or by performing the calculations in a different order to be certain that the numbers agree in both trials. The units need to be examined to ensure that all equations are dimensionally correct. And finally, the answer must be examined to see if it makes sense. An experienced engineer will generally have a good idea of the order of magnitude to expect.

If the answer doesn't seem reasonable, there is probably an error in the mathematics, in the assumptions, or perhaps in the theory used. Judgment is critical. For example, suppose that you are asked to compute the monthly payment required to repay a car loan of $5,000 over a 3-year period at an annual interest rate of 12 percent. Upon solving this problem, you arrived at an answer of $11,000 per month. Even if you are inexperienced in engineering economy, you know that this answer is not reasonable, so you should reexamine your theory and computations. Examination and evaluation of the reasonableness of an answer is a habit that you should strive to acquire. Your instructor and employer alike will find it unacceptable to be given results which you have indicated to be correct but are obviously incorrect by a significant percentage.

The engineering method of problem solving as presented in the previous section is an adaptation of the well-known *scientific problem-solving method*. It is a time-tested approach to problem solving that should become an everyday part of the engineer's thought process. Engineers should follow this logical approach to the solution of any problem while at the same time learning to translate the information accumulated in to a well-documented problem solution.

The following steps parallel the engineering method and provide reasonable documentation of the solution. If these steps are properly executed during the solution of problems in this text and all other courses, it is our belief that you will gradually develop an ability to solve a wide range of complex problems.

1. Problem Statement

State as concisely as possible the problem to be solved. The statement should be a summary of the given information, but it must contain all essential material. Clearly state what is to be determined. For example, find the temperature (K) and pressure (Pa) at the nozzle exit.

2. Diagram

Prepare a diagram (sketch or computer output) with all pertinent dimensions, flow rates, currents, voltages, weights, and so on. A diagram is a very efficient method of showing given and needed information. It also is an appropriate way of illustrating the physical setup, which may be difficult to describe adequately in words. Data that cannot be placed in a diagram should be listed separately.

3. Theory

The theory used should be presented. In some cases, a properly referenced equation with completely defined variables is sufficient. At other times, an extensive theoretical derivation may be necessary because the appropriate theory has to be derived, developed, or modified.

4. Assumptions

Explicitly list in complete detail any and all pertinent assumptions that have been made to realize your solution to the problem. This step is vitally important for the reader's understanding of the solution and its limitations. Steps 3 and 4 might be reversed in some problems.

5. Solution Steps

Show completely all steps taken in obtaining the solution. This is particularly important in an academic situation because your reader, the instructor, must have the means of judging your understanding of the solution technique. Steps completed but not shown make it difficult for evaluation of your work and, therefore, difficult to provide constructive guidance.

6. Identify Results and Verify Accuracy

Clearly identify (double underline) the final answer. Assign proper units. An answer without units (when it should have units) is meaningless. Remember, this final step of the engineering method requires that the answer be examined to determine if it is realistic, so check solution accuracy and, if possible, verify the results.

2.5

Standards of Problem Presentation

Once the problem has been solved and checked, it is necessary to present the solution according to some standard. The standard will vary from school to school and industry to industry.

On most occasions your solution will be presented to other individuals who are technically trained, but you should remember that many times these individuals do not have an intimate knowledge of the problem. However, on other occasions you will be presenting technical information to persons with nontechnical backgrounds. This may require methods different from those used to communicate with other engineers, so it is always important to understand who will be reviewing the material so that the information can be clearly presented.

One characteristic of engineers is their ability to present information with great clarity in a neat, careful manner. In short, the information must be communicated accurately to the reader. (Discussion of drawings or simple sketches will not be included in this chapter, although they are important in many presentations.)

Employers insist on carefully prepared presentations that completely document all work involved in solving the problems. Thorough documentation may be important in the event of legal considerations, for which the details of the work might be introduced into the court proceedings as evidence. Lack of such documentation may result in the loss of a case that might otherwise have been won. Moreover, internal company use of the work is easier and more efficient if all aspects of the work have been carefully documented and substantiated by data and theory.

Each industrial company, consulting firm, governmental agency, and university has established standards for presenting technical information. These standards vary slightly, but all fall into a basic pattern, which we will discuss. Each organization expects its employees to follow its standards. Details can be easily modified in a particular situation once you are familiar with the general pattern that exists in all of these standards.

It is not possible to specify a single problem layout or format that will accommodate all types of engineering solutions. Such a wide variety of solutions exists that the technique used must be adapted to fit the information to be communicated. In all cases, however, one must lay out a given problem in such a fashion that it can be easily grasped by the reader. No matter what technique is used, it must be logical and understandable.

We have listed guidelines for problem presentation. Acceptable layouts for problems in engineering are also illustrated. The guidelines are not intended as a precise format that must be followed, but rather as a suggestion that should be considered and incorporated whenever applicable.

Two methods of problem presentation are typical in the academic and industrial environments. Presentation formats can be either freehand or computer generated. As hardware technology and software developments continue to provide better tools, the use of the computer as a method of problem presentation will continue to increase.

If a formal report, proposal, or presentation is to be the choice of communication, a computer-generated presentation is the correct approach. The example solutions that are illustrated in Figs. 2.1, 2.2, and 2.3 include both freehand as well as computer output. Check with your instructor to determine which method is appropriate for your assignments.

The following 9 general guidelines should be helpful as you develop the freehand skills needed to provide clear and complete problem documentation. The first two examples, Figs. 2.1 and 2.2, are freehand illustrations, and the third example, Fig. 2.3, is computer generated. These guidelines are most applicable to freehand solutions, but many of the ideas and principles apply to computer generation as well.

1. One common type of paper frequently used is called engineering-problems paper. It is ruled horizontally, and vertically on the reverse side, with only heading and margin rulings on the front. The rulings on the reverse side, which are faintly visible through the paper, help one maintain horizontal lines of lettering and provide guides for

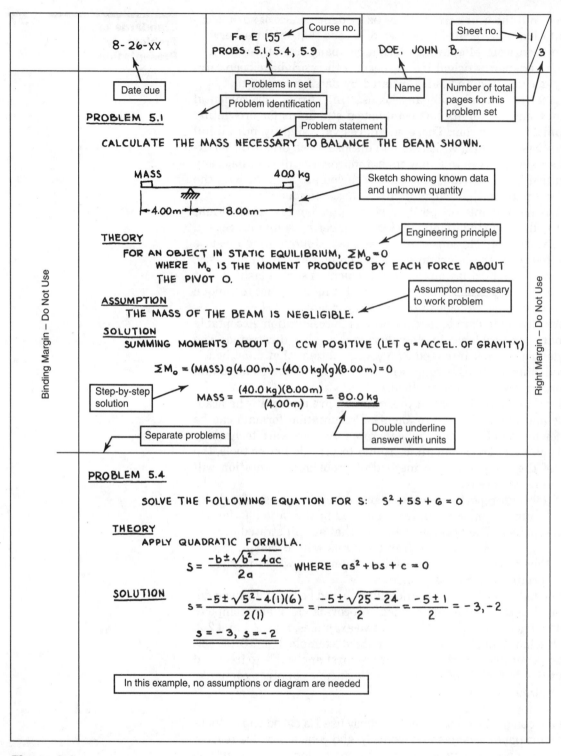

8-26-XX

FR E 155
PROBS. 5.1, 5.4, 5.9

DOE, JOHN B.

Date due

Problems in set

Problem identification

Name

Number of total
pages for this
problem set

PROBLEM 5.1

Problem statement

CALCULATE THE MASS NECESSARY TO BALANCE THE BEAM SHOWN.

MASS   40.0 kg

Sketch showing known data
and unknown quantity

|←4.00m→|←——8.00m——→|

THEORY

Engineering principle

FOR AN OBJECT IN STATIC EQUILIBRIUM, $\Sigma M_o = 0$
WHERE M_o IS THE MOMENT PRODUCED BY EACH FORCE ABOUT
THE PIVOT O.

ASSUMPTION

Assumpton necessary
to work problem

THE MASS OF THE BEAM IS NEGLIGIBLE.

SOLUTION

SUMMING MOMENTS ABOUT O, CCW POSITIVE (LET g = ACCEL. OF GRAVITY)

$$\Sigma M_o = (MASS)\,g\,(4.00m) - (40.0\,kg)(g)(8.00m) = 0$$

Step-by-step
solution

$$MASS = \frac{(40.0\,kg)(8.00m)}{(4.00m)} = \underline{\underline{80.0\,kg}}$$

Separate problems

Double underline
answer with units

PROBLEM 5.4

SOLVE THE FOLLOWING EQUATION FOR S: $S^2 + 5S + 6 = 0$

THEORY

APPLY QUADRATIC FORMULA.

$$S = \frac{-b \pm \sqrt{b^2 - 4ac}}{2a} \text{ WHERE } as^2 + bs + c = 0$$

SOLUTION

$$s = \frac{-5 \pm \sqrt{5^2 - 4(1)(6)}}{2(1)} = \frac{-5 \pm \sqrt{25-24}}{2} = \frac{-5 \pm 1}{2} = -3, -2$$

$$\underline{\underline{s = -3,\ s = -2}}$$

In this example, no assumptions or diagram are needed

Figure 2.1
Elements of a problem layout.

<u>PROBLEM 13.1</u>   SOLVE FOR THE VALUE OF RESISTANCE R
IN THE CIRCUIT SHOWN BELOW.

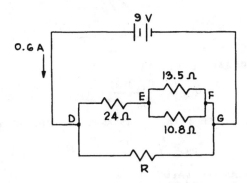

<u>THEORY</u>

- FOR RESISTANCES IN PARALLEL: $\dfrac{1}{R_{TOTAL}} = \dfrac{1}{R_1} + \dfrac{1}{R_2} + \dfrac{1}{R_3} + \ldots$

  THUS FOR 2 RESISTANCES IN PARALLEL

  $$R_{TOTAL} = \frac{R_1 R_2}{R_1 + R_2}$$

- FOR RESISTANCES IN SERIES: $R_{TOTAL} = R_1 + R_2 + R_3 + \ldots$

- OHM'S LAW: $E = RI$ WHERE E = ELECT. POTENTIAL IN VOLTS
  I = CURRENT IN AMPERES
  R = RESISTANCE IN OHMS

<u>SOLUTION</u>

- CALCULATE EQUIVALENT RESISTANCE BETWEEN POINTS E AND F.
  RESISTORS ARE IN PARALLEL.

  $$\therefore R_{EF} = \frac{R_1 R_2}{R_1 + R_2} = \frac{(13.5)(10.8)}{13.5 + 10.8} = \frac{145.8}{24.3} = 6.00 \ \Omega$$

- CALCULATE EQUIVALENT RESISTANCE OF UPPER LEG
  BETWEEN D AND G.

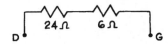

  SERIES CIRCUIT

  $$\therefore R'_{DG} = R_{24} + R_6 = 24 + 6 = 30 \Omega$$

In this example, no assumptions were necessary

Figure 2.2
Sample problem presentation.

59

- CALCULATE EQUIVALENT RESISTANCE BETWEEN D AND G.

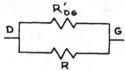

PARALLEL RESISTORS, SO

$$R_{DG} = \frac{(R'_{DG})(R)}{R'_{DG} + R}$$

- CALCULATE TOTAL RESISTANCE OF CIRCUIT USING OHM'S LAW.

$$R_{DG} = \frac{E}{I} = \frac{9V}{0.6A} = 15 \ \Omega$$

- CALCULATE VALUE OF R
  FROM PREVIOUS EQUATIONS.

$$R_{DG} = 15 \ \Omega = \frac{(R'_{DG})(R)}{R'_{DG} + R} = \frac{(30)(R)}{30 + R}$$

SOLVING FOR R:

$$(30 + R)(15) = 30R$$

$$30 + R = 2R$$

$$\underline{R = 30 \ \Omega}$$

Figure 2.2 (cont.)

60

Date	Engineering	Name:_____

Problem

A tank is to be constructed that will hold 5.00×10^5 L when filled. The shape is to be cylindrical, with a hemispherical top. Costs to construct the cylindrical portion will be $300/$m^2$, while costs for the hemispherical portion are slightly higher at 400 /m^2.

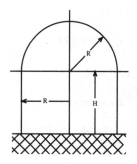

Find

Calculate the tank dimensions that will result in the lowest dollar cost.

Theory

Volume of cylinder is... $\qquad V_c = \pi R^2 H$

Volume of hemisphere is... $\qquad V_H = \dfrac{2\pi R^3}{3}$

Surface area of cylinder is... $\qquad SA_c = 2\pi R H$

Surface area of hemisphere is... $\qquad SA_H = 2\pi R^2$

Assumptions

Tank contains no dead air space
Construction costs are independent of size
Concrete slab with hermetic seal is provided for the base.
Cost of the base does not change appreciably with tank dimensions.

Solution

1. Express total volume in meters as a function of height and radius

$$V_{Tank} = f(H, R)$$
$$= V_C + V_H$$
$$500 = \pi R^2 H + \frac{2\pi R^3}{3}$$

Note: $\quad 1 m^3 = 1000L$

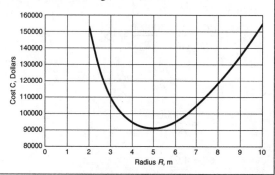

Figure 2.3
Sample problem presentation.

61

2. Express cost in dollars as a function of height and radius

$$C = C(H, R)$$

$$= 300\,(SA_C) + 400\,(SA_H)$$

$$= 300\,(2\pi RH) + 400\,(2\pi R^2)$$

Note: Cost figures are exact numbers

3. From part #1 solve for $H = H(R)$

$$H = \frac{500}{\pi R^2} - \frac{2R}{3}$$

4. Solve cost equation, substituting $H = H(R)$

$$C = 300\left[2\pi R\left(\frac{500}{HR^2} - \frac{2R}{3}\right)\right] + 400\left(2\pi R^2\right)$$

$$C = \frac{300000}{R} + 400\pi R^2$$

5. Develop a table of Cost vs. Radius and plot graph.

6. From graph select minimum cost.

R = <u>5.00m</u>
C = $91 000

7. Calculate H from part 3 above

H = <u>3.033 m</u>

8. Verification / check of results from the calculus:

$$\frac{dC}{dR} = \frac{d}{dR}\left[\frac{300000}{R} + 400\pi R^2\right]$$

$$= \frac{-300000}{R^2} + 800\pi R = 0$$

$$R^3 = \frac{300000}{800\pi}$$

R = <u>4.92m</u>

Cost vs. Radius

Radius R, m	Cost C, $
1.0	301 257
2.0	155 027
3.0	111 310
4.0	95 106
5.0	91 416
6.0	95 239
7.0	104 432
8.0	117 925
9.0	135 121
10.0	155 664

Figure 2.3 (cont.)

sketching and simple graph construction. Moreover, the lines on the back of the paper will not be lost as a result of erasures.

2. The completed top heading of the problems paper should include such information as name, date, course number, and sheet number. The upper right-hand block should normally contain a notation such as a/b, where a is the page number of the sheet and b is the total number of sheets in the set.

3. Work should ordinarily be done in pencil using an appropriate lead hardness (F or H) so that the linework is crisp and unsmudged. Erasures should always be complete, with all eraser particles removed. Letters and numbers must be dark enough to ensure legibility when photocopies are needed.

4. Either vertical or slant letters may be selected as long as they are not mixed. Care should be taken to produce good, legible lettering but without such care that little work is accomplished.

5. Spelling should be checked for correctness. There is no reasonable excuse for incorrect spelling in a properly done problem solution.

6. Work must be easy to follow and uncrowded. This practice contributes greatly to readability and ease of interpretation.

7. If several problems are included in a set, they must be distinctly separated, usually by a horizontal line drawn completely across the page between problems. Never begin a second problem on the same page if it cannot be completed there. It is usually better to begin each problem on a fresh sheet, except in cases where two or more problems can be completed on one sheet. It is not necessary to use a horizontal separation line if the next problem in a series begins at the top of a new page.

8. Diagrams that are an essential part of a problem presentation should be clear and understandable. Students should strive for neatness, which is a mark of a professional. Often a good sketch is adequate, but using a straightedge can greatly improve the appearance and accuracy of a diagram. A little effort in preparing a sketch to approximate scale can pay great dividends when it is necessary to judge the reasonableness of an answer, particularly if the answer is a physical dimension that can be seen on the sketch.

9. The proper use of symbols is always important, particularly when the International System (SI) of units is used. It involves a strict set of rules that must be followed so that absolutely no confusion of meaning can result. There are also symbols in common and accepted use for engineering quantities that can be found in most engineering handbooks. These symbols should be used whenever possible. It is important that symbols be consistent throughout a solution and that all are defined for the benefit of the reader and for your own reference.

The physical layout of a problem solution logically follows steps similar to those of the engineering method. You should attempt to present the process by which the problem was solved in addition to the solution so that any reader can readily understand all aspects of the solution. Figure 2.1 illustrates the placement of the information.

Figures 2.2 and 2.3 are examples of typical engineering-problem solutions. You may find that they are helpful guides as you prepare your problem presentations.

2.6

Key Terms and Concepts

The following terms are basic to the material in Chapter 2. You should be able to define these terms and to be able to interpret them into various applications.

Process   Problem presentation

Analysis   Solution documentation

Engineering method

Problems

The solution to lengths and angles of oblique triangles can be arrived at by application of fundamental trigonometry. All angles are to be considered precise numbers. Solve the following problems using Fig. 2.4 as a general guide.

Figure 2.4

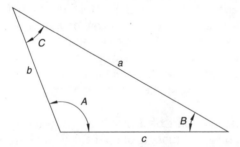

2.1 Given one side and two angles of an oblique triangle.

$C = 75°$ $a = 1\,255$ m

$A = 30°$

Using laws of sines and sum of angles, determine the angle B and distances b and c.

2.2 Given two sides and the included angle of an oblique triangle.

$C = 40°$ $a = 75$ in

$b = 44$ in

Using the law of cosines and sum of angles, determine the angles A and B and the distance c.

2.3 Given the sides of an oblique triangle.

$a = 440$ ft

$b = 910$ ft

$c = 1\,285$ ft

Using the law of cosines and sum of angles, determine the angles A, B, and C.

2.4 Given two sides and the included angle of an oblique triangle.

$A = 20°$ $\qquad b = 2\,550$ m

$\qquad\qquad c = 1\,825$ m

(a) Using the sum of angles, law of sines, and law of tangents, determine the angles B and C and the distance a.

(b) Compute the radius (r), of an inscribed circle and the radius (R) of the circumscribed circle.

Vector quantities have both magnitude and direction. Figure 2.5 illustrates convention for Probs. 2.5 and 2.6, with positive angles measured clockwise from the $+$ x-axis.

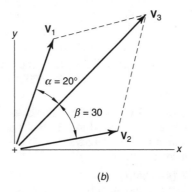

Figure 2.5

(a) (b)

2.5 A vector **A** has a magnitude of 325 m at $\theta = -52°$ (ccw) from the $+$ x-axis.

(a) Determine the magnitude of $|Ax|$ and $|Ay|$.

(b) Determine **A** (magnitude and direction) if $Ax = 85$ m and $Ay = 33$ m.

2.6 A wind vector V_3 is the sum of vectors V_1 and V_2. Vector V_2 makes angle of -15 with the $+$ x-axis and has a magnitude of 20.0 mph. Determine the magnitude and direction of vector V_3.

2.7 A survey crew determined the angles and distances given in Fig. 2.6. Calculate the distance across the lake at DE.

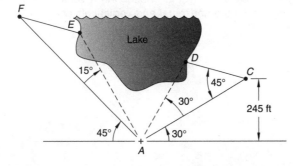

Figure 2.6

2.8 An engineering student in a stationary hot-air balloon is momentarily fixed at 1,325-ft elevation above a level piece of land. The pilot looks down (60° from horizontal) and turns laterally 360°. How many acres are contained within the core generated by his line of sight? How high would the balloon be if, when performing the same procedure, an area 4 times greater is encompassed?

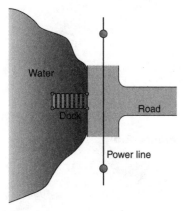

Figure 2.7

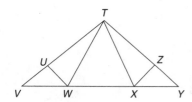

2.9 A small aircraft has a glide ratio of 15:1. (This glide ratio means the plane moves 15 units horizontally for each 1 unit in elevation.) You are exactly in the middle of a 3.0 mile diameter lake at 5.00×10^2 ft when the fuel supply is exhausted. You see a gravel road by the dock (Fig. 2.7) and must decide whether to try a ground or water landing. Show appropriate assumptions and calculations to support your decision.

2.10 A pilot in an ultralight knows that her aircraft in landing configuration will glide 2.0×10^1 km from a height of 2.0×10^3 m. A TV transmitting tower is located in a direct line with the local runway. If the pilot glides over the tower with 3.0×10^1 m to spare and touches down on the runway at a point 6.5 km from the base of the tower, how high is the tower?

2.11 A simple roof truss design is shown in Fig. 2.8. The lower section VWXY is made from three equal-length segments. UW and XZ are perpendicular to VT and TY, respectively. If VWXY is 24 m, and the height of the truss is 12 m, determine the lengths of XT and XZ.

2.12 If you are traveling at 65 mph, what is your angular axle speed in RPM? Assume a tire size of P235/75R15, which has an approximate diameter of 30.0 in.

2.13 The wheel of an automobile turns at the rate of 195 RPM. Express this angular speed in (a) revolutions per second and (b) radians per second. If the wheel has a 30.0-in diameter, what is the velocity of the auto in miles per hour?

2.14 A bicycle wheel has a 28.0-in diameter wheel and is rotating 195 RPM. Express this angular speed in radians per second. How far will the bicycle travel (miles) in 45.0 minutes and what will be the velocity in miles per hour?

2.15 Assume the earth's orbit to be circular at 93.0×10^6 mi about the sun. Determine the speed of the earth (in miles per second) around the sun if there are exactly 365 days per year.

2.16 A child swinging on a tree rope 8.00 m long reaches a point 4.00 m above the lowest point. Through what total arc in degrees has the child passed? What distance in meters has the child swung?

2.17 Two engineering students were assigned the job of measuring the height of an inaccessible cliff (Fig. 2.9). The angles and distances shown were measured on a level beach in a vertical plane due south of the cliff. Determine the horizontal and vertical distances from A to B.

Figure 2.8

Figure 2.9

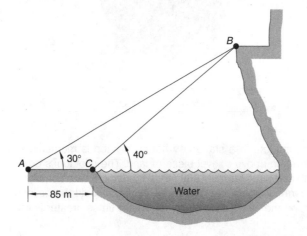

2.18 A survey team with appropriate equipment has been asked to measure a nonrectangular plot of land *ABCD* (Fig. 2.10). The following data were recorded: $CD = 165.0$ m, $DA = 180.0$ m, and $AB = 115.0$ m. Angle $DAB = 120$ and angle $DCB = 100$.
   (*a*) Calculate the length of side *BC* and the area of the plot.
   (*b*) Estimate the water surface area within the plot and list assumptions.
   (*c*) Using answers from part b, what percentage of the total surface area within *ABCD* is land?

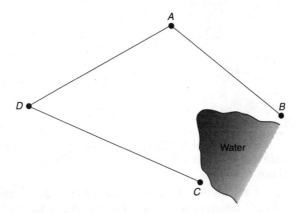

Figure 2.10

2.19 In Fig. 2.11 points *S*, *T*, *U*, and *V* are survey markers in a land development located on level terrain. The distance *ST* measures 355.0 ft. Angles at each marker were recorded as follows: $STU = 80°$, $TUS = 70°$, $VUS = 50°$, and $UVS = 70°$.
   (*a*) Calculate the distances *TU*, *UV*, *VS*, and *SU*.
   (*b*) Determine the area of *STUV* in acres.

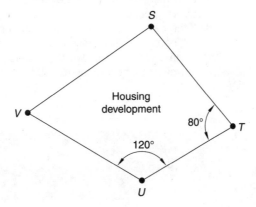

Figure 2.11

2.20 Circular sheets of metal 70.0 cm in diameter are to be used for stamping new highway signs. Calculate the percent of waste when the largest possible inscribed shape is:
   (*a*) An equilateral triangle
   (*b*) A square
   (*c*) An octagon

2.21 Three circles are tangent to each other as in Fig. 2.12. The respective radii are 1.65×10^3, 10.00×10^2, and 7.75×10^2 mm.

(a) Find the area of the triangle (in square millimeters) formed by joining the 3 centers.

(b) Determine the area within the triangle that is outside of the circles.

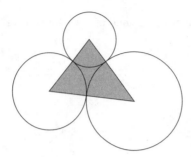

Figure 2.12

2.22 A waterwheel turns a belt on a drive wheel for a flour mill. The pulley on the waterwheel is 2.00 m in diameter, and the drive pulley is 0.500 m in diameter. If the centers of the pulleys are 4.00 m apart, calculate the length of belt needed.

2.23 A narrow, flat belt is used to drive a 50.00 cm (diameter) pulley from a 10.00 cm (diameter) pulley. The centers of the 2 pulleys are 50.00 cm apart. How long must the belt be if the pulleys rotate in the same direction? In opposite directions?

2.24 A homeowner decides to install a family swimming pool. It is 32.0 ft in diameter with a 4.00 ft water level. The cross-section of the pool is illustrated in Fig. 2.13. Consider the following problems.

(a) How many cubic feet of soil must be excavated to accommodate the pool liner if the bottom profile is as illustrated in Fig. 2.13?

(b) How many gallons of water will be required to fill the pool to 4 ft above ground?

(c) If the owner moves the water from a nearby lake, how many tons will be carried? (Density of water is 62.4 lbm/ft^3.)

(d) If the owner carries two 5-gal pails per trip and makes 20 trips per evening, how many days will it take to fill the pool?

Figure 2.13

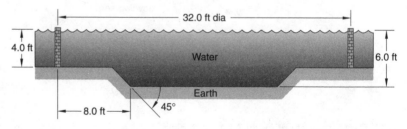

2.25 A block of metal has a 90° notch cut from its lower surface. The notched part rests on a circular cylinder 4.0 in in diameter as shown in Fig. 2.14. If the lower surface of the block is 2.5 in above the base plate, how deep is the notch?

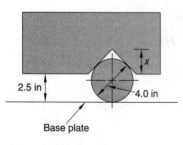

Base plate

Figure 2.14

2.26 Show that the area of the shaded segment in Fig. 2.15 is given by the expression

$$As = \frac{r^2}{2}(\phi - \sin \phi)$$

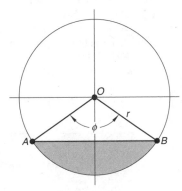

Figure 2.15

2.27 A fuel tank is 10.00 ft in diameter and 20.00 ft long. The tank is buried vertically as shown in Fig. 2.16. Develop an expression for volume (V) in gallons for any depth (h) in feet for this configuration.

2.28 Consider the fuel tank in Prob. 2.27 positioned on its side instead of its base. The tank length is 30.0 ft.

(a) Develop an expression for volume (V) in gallons as a function of radius (r) in feet and height (h) in feet.

(b) If the empty tank with radius 5.00 ft has a mass of 15.00×10^2 lbm, develop an expression for the mass of tank and fuel (in pounds-mass) as a function of height (h) in feet. Density of the fuel is 815 kg/m^3.

(c) How many gallons of fuel are in the tank at a height (h) of 4.50 ft? At that depth, what is the total mass of tank and fuel?

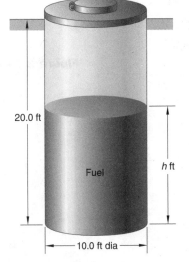

Figure 2.16

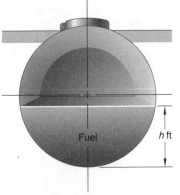

*End view of buried cylindrical tank.

Figure 2.17

2.29 Eighteen circular wooden bases are to be cut from a piece of 3/4-in plywood. Each circular base has a diameter of 10.00 in. Assume a negligible saw blade thickness for the following problems.

    (a) What is the area of triangle *ABC*?

    (b) Calculate the area of waste material between *X* and *Y* above *BC*.

    (c) Determine the dimensions of the rectangular piece of plywood shown in Fig. 2.18 to the nearest 0.5 in.

    (d) What is the area of the largest piece of waste material?

    (e) What is the percentage of waste given the configuration shown in Fig. 2.18?

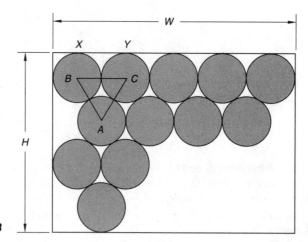

Figure 2.18

Engineering Design—A Process

Engineering design is a systematic process by which solutions to the needs of humankind are obtained. The process is applied to problems (needs) of varying complexity. For example, mechanical engineers will use the design process to find an effective, efficient method to convert reciprocating motion to circular motion for the drive train in an internal combustion engine; electrical engineers will use the process to design electrical generating systems using falling water as the power source; and materials engineers use the process to design high-temperature materials which enable astronauts to safely reenter the earth's atmosphere.

The vast majority of complex problems in today's high-technology society depend for solution not on a single engineering discipline but, rather, on teams of engineers, scientists, environmentalists, economists, sociologists, and legal personnel. Solutions are not only dependent on the appropriate applications of technology but also on public sentiment as executed through government regulations and political influence. As engineers we are empowered with the technical expertise to develop new and improved products and systems, but at the same time we must be increasingly aware of the impact of our actions on society and the environment in general and work conscientiously toward the best solution in view of all relevant factors.

Design is the culmination of the engineering educational process; it is the salient feature that distinguishes engineering from other professions.

A formal definition of *engineering design* is found in the curriculum guidelines of the Accreditation Board for Engineering and Technology (ABET). ABET accredits curricula in engineering schools and derives its membership from the various engineering professional societies. Each accredited curriculum has a well-defined design component which falls within the ABET guidelines. The ABET statement on design reads as follows:

Engineering design is the process of devising a system, component, or process to meet desired needs. It is a decision-making process (often iterative), in which the basic sciences, mathematics, and engineering sciences are applied to convert resources optimally to meet a stated objective. Among the fundamental elements of the design process are the establishment of objectives and criteria, synthesis, analysis, construction, testing, and evaluation. The engineering design component of a curriculum must include most of the following features: development of student creativity, use of open-ended problems, development and use of modern design theory and methodology, formulation of design problem statements and specifications, consideration of alternative solutions, feasibility considerations, production processes, concurrent engineering design, and detailed system descriptions. Further, it is essential to include a variety of realistic constraints such as economic factors, safety, reliability, aesthetics, ethics, and social impact.

In order for you to gain the fundamental knowledge and experience needed to understand the design process, you must partake in meaningful design activities as a student. To assist you in your first activity, we will take you through an actual student design project as undertaken by a team of beginning engineering students. The project is guided by the 10-step design process listed in Fig. 3.1. As you study the design process and the description of how the student team accomplished each step, pay particular attention to the structure of the process rather than to the development of the particular solution arrived at by the team. By doing this you will understand how engineers approach a need, develop alternative solutions, select the best solution, and communicate the results.

Throughout this chapter you will see many examples of the utilization of computers in engineering design (see Fig. 3.2). The computer is a major tool for the engineer in the acquisition and analysis of data, and the definition of potential solutions. Throughout history engineers have used the best computational devices available at the time to obtain solutions to problems. The computer is the fastest and most powerful computational tool yet conceived; it produces numerical computations at heretofore unheard of rates. It also provides an insight to problems and solutions through its capability to simulate actual phenomena. This capability provides the engineer a tremendous advantage in developing new and improved products in a much shorter time frame than ever before. The rapidly developing computer envi-

Figure 3.1
A 10-step design process.

1. Identification of a need
2. Problem definition
3. Search
4. Constraints
5. Criteria
6. Alternative solutions
7. Analysis
8. Decision
9. Specification
10. Communication

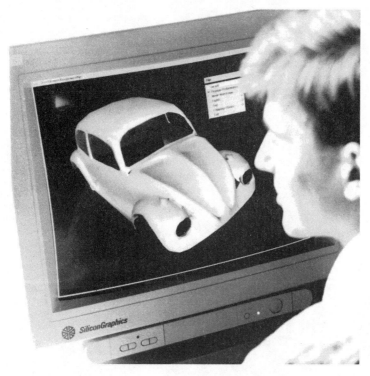

Figure 3.2
A design team member views a three-dimensional model created on an engineering workstation.

ronment will enable you to be a better educated and more productive engineer than your predecessors.

3.1.1
The Design Process

A simple definition of *design* is "a structured problem-solving activity." A *process,* on the other hand, is a phenomenon identified

Figure 3.3
A student team reviews a part drawing for a design project with the instructor.

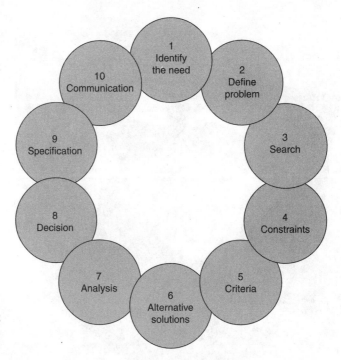

Figure 3.4
The design process is iterative
in nature.

through step-by-step changes that lead toward a required result. Both these definitions suggest the idea of an orderly, systematic approach to a desired end. Figure 3.4 shows the design process approach as continuous and cyclic in nature. This idea has validity in that many problems arise during the design process that generate subsequent designs. You should not assume that each of your design experiences will necessarily follow the sequential steps without deviation. Experienced designers will agree that the steps as shown are quite logical; but on many occasions designers have had to repeat some steps or perhaps have been able to skip one or more steps.

Before beginning an overview of the entire design process, we must state that limits are always placed on the amount of time available. Normally, we establish a time frame or a series of deadlines for ourselves before we begin the process. It is almost impossible for us to tell you how much time should be allocated for each step because the problems are so varied. A sample of a time frame is shown in Fig. 3.5.

The whole process begins when a need is recognized: In essence, a human need has been identified. Often it is not the designers who are involved in step 1, but they usually assist in defining the problem (step 2) in terms that allow it to be scrutinized. Information is gathered in step 3, and then boundary conditions (constraints) are established (step 4). The criteria against which the alternative solutions are compared are chosen in step 5. At step 6 several possible solutions are entertained and the creative, innovative talents of the designer come into play. This is followed

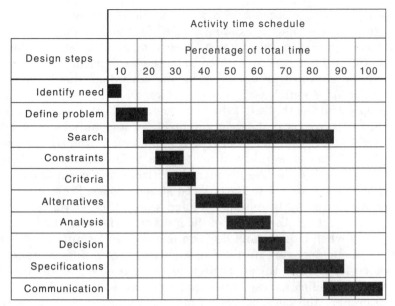

Design steps	Activity time schedule									
	Percentage of total time									
	10	20	30	40	50	60	70	80	90	100
Identify need	█									
Define problem	█	█								
Search		█	█	█	█	█	█	█	█	
Constraints			█							
Criteria			█							
Alternatives				█	█					
Analysis						█				
Decision						█				
Specifications								█	█	
Communication									█	█

Figure 3.5
In order to control the design
process, a time schedule must
be developed early.

by detailed analysis of the alternatives (step 7), after which a decision is reached regarding which one should be completely developed (step 8). Specifications of the chosen concept are prepared (step 9), and its merits are explained to the proper people or agencies (step 10) so that implementation (construction, production, and so on) can be accomplished. A more detailed explanation of each step, as well as a reporting of the actions of the student design team, constitutes the remainder of this chapter.

3.2

Identification of a Need—Step 1

Before the process can begin, someone has to recognize that some constructive action needs to be initiated. This may sound vague, but understandably so because such is the way the process normally begins. Engineers do not have supervisors who tell them to come to work tomorrow and "identify a need." You might be asked to do so in the classroom because some professors may have you work on a project that you choose rather than one that is assigned. When most of us speak of a need, we generally refer to a lack or shortage of something we consider essential or highly desirable. Obviously, this is an extremely relative thing for what may be a necessity to some could be a luxury to others.

More often than not, then, someone other than the engineer decides that a need exists. In industry, or the private sector, it is essential that products sell for the company to survive. Most of the products have a life cycle that goes from the development stage, when the expenditures by the organization are high and sales are low, to the peak demand period, when profits are high, and eventually to the point where the product becomes obsolete. Even though a human need may still exist, the economic demand does not because a more attractive alternative has become available. With obsolescence of a product, the company perceives a need to phase out the existing product and to develop a new one that is profitable. Inasmuch as most companies exist to make a profit, profit can be considered to be the basic need.

A bias toward profit and economic advantage should not be viewed as a selfish position because products are purchased by people who feel that what they are buying will satisfy a need that they perceive as real. Society appreciates anyone who provides essential and desirable services, as well as goods that we use and enjoy. The consumers are ultimately the judges of whether there is truly a need. In like manner, the citizens of a community decide whether or not to have paved streets, parks, libraries, adequate police and fire protection, and scores of other things. City councils vote on the details of the programs. However, during the period when citizens and decision makers are formulating their plans, engineers are involved in supplying factual information to assist them. After the policy decisions have been made, engineers conduct studies, surveys, tests, and computations that allow them to prepare the detailed design plans, drawings, etc. that shape the final project.

3.2.1
The Chapter Example–Step 1 of the Design Process

Throughout the remainder of the chapter we will trace the steps that five beginning engineering students* followed to produce a design for their class project. As a starting point, a professor may assign students the task of identifying a need. It usually is easier to approach such an assignment by beginning with a very broad area of technology, such as energy.

The five students who were chosen were informed that all the student teams would be involved in some area dealing with the energy problem. Their professor began with construction of a decision tree, shown in Fig. 3.6. The class discussed sources of energy and jointly added the first level of subproblems: fossil, wind, geothermal, solar, nuclear, and organic. The class was then divided into groups; the groups began to subdivide further *one* of the energy sources listed above. There probably is no end to this procedure, but it does provide quickly a wide range of topics from which needs may be more easily recognized. Our student design team developed the organic section of Fig. 3.6, as shown, and thus began their discussion of the general topic of firewood for use in fireplaces. They recorded statements such as the following:

1. A large number of people have decided to install fireplaces in their homes and apartments.

*The five students were John W. Benike, Douglas L. Carper, Patrick J. Grablin, Rick Sessions, and David L. White.

2. Firewood is not as commercially available as it used to be.

3. The price of firewood has risen significantly.

4. People are now more willing to cut and split their own firewood than they were previously.

5. The small, inexpensive chain saw has made the cutting portion of the task more acceptable, but splitting the wood is still a major problem.

After a reasonable period of discussion involving these topics as well as others related to firewood, they agreed upon the following initial statement of need:

> There is a need for an inexpensive supply of firewood for use in the home.

(We will see later that this statement was changed by the students in much the same way that professional engineers refine and redefine problems during a design process.)

Figure 3.6
A decision tree pertaining to energy.

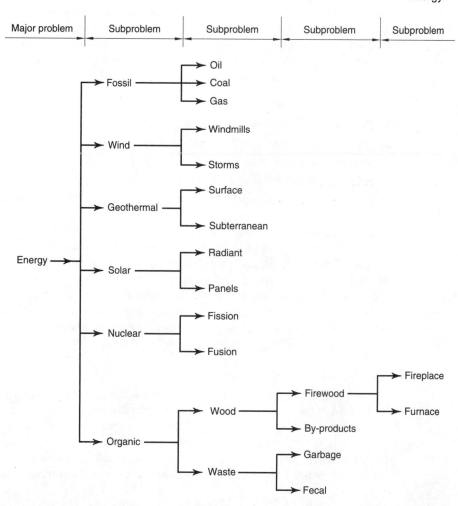

Problem Definition—Step 2

There is often a temptation to construct quickly a mental picture of a gadget that if properly designed and manufactured will satisfy a need. In the general case of a need for firewood we obviously know that we can call a supplier and have firewood delivered, but the cost factor is equally obvious. We could get a bit facetious and decide to burn our furniture. There have been emergency situations during extreme storms when this was the best solution available. If some friendly neighbor will supply us with firewood at no cost or effort, a problem does not exist. An important point to realize at this time is that had we allowed ourselves to focus on a specific piece of equipment or a single method of obtaining firewood in the very beginning, we would never have considered the statements mentioned earlier.

3.3.1
Broad Definition First

The need as previously stated does not point to any particular solution and thereby leaves us with the opportunity to consider a wide range of alternatives before we agree on a specific problem statement. Consider for a moment a partial array of possibilities that will satisfy the original statement that there is a need for an inexpensive supply of firewood for use in the home:

1. Purchase firewood from a supplier.
2. Purchase standing timber to be cut for firewood.
3. Purchase trees that have been cut or that have fallen.
4. Hire a portion of the work done (probably the splitting of logs).
5. Design improved equipment to facilitate a portion of the process.

This is not an exhaustive list. You may want to take a few minutes to add to it. The first item is the current solution for many people. The other items show some promise of being an improvement over simply purchasing the firewood. You may think that item 1 should not be listed because it offers no change. If so, you are mistaken because the status quo is the solution that is selected most often, at least as a temporary measure.

3.3.2
Symptom versus Cause

If you cough and do nothing but suck on a cough drop, you may be treating the symptom (the tickle in your throat) but doing little to alleviate the cause of the tickle. This approach may be expedient; however, it can result many times in a repetition of the problem if the tickle is caused by a virus or a foreign object of some sort. Engineers seldom tell a client to take two aspirins and call back tomorrow, but they can sometimes be guilty of failing to see the real or root problem.

For many years residential subdivisions were designed so that the rainfall would drain away quickly, and expensive storm sewer systems were constructed to accomplish this task. Not only were the sewers expensive, but they also resulted in transporting the water problem downstream for someone else to handle. In recent years perceptive engineers have designed land developments so that the rainfall is temporarily collected in "holding pools" and released gradually over a longer period of time. This approach employs smaller, less expensive sewers and reduces the likelihood of flooding downstream. The real problem was not how to get rid of the rainfall as rapidly as possible, but how to control the water.

3.3.3
Solving the Wrong Problem

In the 1970s the problem of increasing fatalities as a result of auto accidents was clearly recognized. It was shown that the fatality rate could be significantly reduced if the driver and front seat passenger used lap and shoulder belts. The solution technique that was implemented was to build in an interlock system that required the belts to be latched before the auto could be started. That solution certainly should have solved the problem but it did not. It attacked the problem of requiring that the belts be physically used, but it did nothing to solve the real problem—the driver's attitude. The driver and passenger still did not wish to use belts and did everything possible to avoid it, even by having the interlock system removed.

3.3.4
The Chapter Example—Step 2

The students whose progress we are following considered the possibilities outlined in Sec. 3.3.1. Their discussions covered a range of topics, and from their notebook we have listed a few of their pertinent recorded thoughts:

1. The range of possible solutions must be reduced before the problem can be solved.

2. People really like the smell of burning wood.

3. An adequate supply of wood already exists in most areas.

4. Chain saws are already well developed; hence, cutting the wood into proper lengths is not a pertinent problem.

5. Time is very short, so a direction must be selected that we (the students) can achieve.

Most assuredly these young people had other thoughts, many of which were not recorded. The result of their consideration was a slightly revised problem definition:

There needs to be available to the average household an inexpensive, efficient method of splitting a small quantity of firewood.

In practice, you and other engineers will face restrictions that will affect the quality of your solutions. Many times your solution will have to meet governmental regulations in order to qualify for grants of money; or perhaps safety requirements by some agency cannot be met if certain materials are used. In almost all of your projects, there will be cost and time constraints that force you to make decisions that are not what you really want to do. Such decisions, once made, then control many of your subsequent actions on that project.

This situation was faced by our student design team. By limiting the range of possible solutions and accepting the present method of cutting wood, they eliminated even the consideration of other burning materials. We are not being critical of their decision because we have experienced similar time and resource constraints.

3.4
Search—Step 3

Most of your productive professional time will be spent locating, applying, and transferring information—all sorts of information. This is not the popular opinion of what engineers do, but it is the way it will be for you. Engineers are problem solvers, skilled in applied mathematics and science, but they seldom, if ever, have enough information about a problem to begin solving it without first gathering more data. This search for information may reveal facts about the situation that result in redefinition of the problem (see Fig. 3.7).

3.4.1
Types of Information

The problem usually dictates what types of data are going to be required. The one who recognizes that something was needed (step 1) probably listed some things that are known and some things that need to be known. The one or ones who defined the problem had to have knowledge of the topic or they could not have done their part (step 2). Generally, there are several things that we look for in beginning to solve most problems. For example,

1. What has been written about it?
2. Is something already on the market that may solve the problem?
3. What is wrong with the way it is being done?
4. What is right with the way it is being done?
5. Who manufactures the current "solution"?
6. How much does it cost?
7. Will people pay for a better one if it costs more?
8. How much will they pay (or how bad is the problem)?

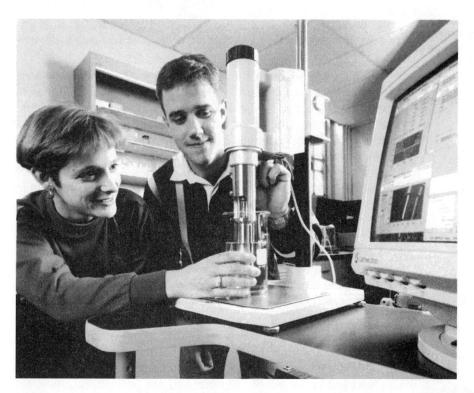

3.4.2
Sources of Information

If anything can be said about the last half of the twentieth century, it is that we have had an explosion of information. The amount of data that can be uncovered on most subjects is overwhelming. People in the upper levels of most organizations have assistants who condense most of the things that they must read, hear, or watch. When you begin a search for information, be prepared to scan many of your sources and document their location so that you can find them easily if the data subsequently appear to be important.

Some sources that are available include the following:

1. *Existing solutions.* Much can be learned from the current status of solutions to a specific need if actual products can be located, studied and, in some cases, purchased for detailed analysis. If a product can be acquired in the marketplace, a process called *reverse engineering* can be performed to determine answers to some questions posed in the section on Types of Information. An improved solution or an innovative new solution cannot be found unless the existing solutions are thoroughly understood. Reverse engineering is an excellent learning technique for students and engineers in industry who are beginning to apply the design process. You should look to local industries and retail outlets for existing solutions that currently satisfy the need you have identified. In some instances you may be able to purchase the product or at least observe a demonstration of its capability.

2. *Internet.* The electronic Internet that today connects millions of computer users in nearly 100 nations to a global "information superhighway" provides rapid access to a wealth of knowledge. Industries, businesses, government organizations, and educational institutions are now connected to the Internet enabling nearly instantaneous access to research data, product information, special interest groups, and experts throughout the world. Although you may not yet have access to the Internet from your personal computer, your university has facilities from which you can gain access. The World Wide Web (WWW) is an Internet-based navigational system which provides the capability to find information and move to different locations (servers) with point and click ease. The WWW provides access to graphic images, photographs, audio, and full-motion video. Accessing the Internet for information should be a major component in any research effort.

3. *Your library.* Many universities have courses that teach you how to use your library. Such courses are easy when you compare them with those in chemistry and calculus, but their importance should not be underestimated. There are many sources in the library that can lead you to the information that you are seeking. You may find what you need in an index such as the *Engineering Index,* but do not overlook the possibility that a general index, such as *The Reader's Guide* or *Business Periodicals Index,* may also be useful. The *Thomas Register of American Manufacturers* may direct you to a company that makes a product that you need to know more about. *Sweets Catalog* is a compilation of manufacturer's information sheets and advertising material about a wide range of products. There are many other indexes that provide specialized information. The nature of your problem will direct which ones may be helpful to you. Do not hesitate to ask for assistance from the librarian. You should use to advantage the computer databases found in libraries and often available through CD-ROM technology.

4. *Government documents.* Many of these are housed in special sections of your library, but others are kept in centers of government—city halls, county court houses, state capitols, and Washington, D.C. The regulatory agencies of government, such as the Interstate Commerce Commission, the Environmental Protection Agency, and regional planning agencies, make rules and police them. The nature of the problem will dictate which of the myriad of agencies can fill your needs.

5. *Professional organizations.* The American Society of Civil Engineers is a technical society that will be of interest to students majoring in civil engineering. Each major in your college is associated with not one but often several such societies. The National Society of Professional Engineers is an organization that most engineering students will eventually join, as well as at least one technical society such as the American Society of Mechanical Engineers (ASME), the Institute of Electrical and Electronics Engineers (IEEE), or any one of dozens that serve the technical interests of the host of specialties with which professional practices seem most closely associated. Many engineers are members of several associations and societies. Other organizations, such as the American Medical Association and the American Bar Association, serve various professions, and all have publications and referral services.

6. *Trade journals.* They are published by the hundreds, usually specializing in certain classes of products and services.

7. *Vendor catalogs.* Perhaps your college subscribes to one of the several information services that gather and index journals and catalogs. These data banks may have tens of thousands of such items available to you on microfilm. You need only learn how to use them.

8. Individuals that you have reason to believe are experts in the field. Your college faculty has at least several such individuals, maybe many. There are, no doubt, some practicing engineers in your city.

3.4.3
Recording your Findings

The purpose of a bibliography is to direct you to more information than is included in the article you are reading. The form of the bibliography makes it easy to find the reference. So it seems reasonable for you to record your information sources in proper form so that if that reference is to be cited in your report, you are ready to do it properly. By so doing, you are ensuring that it can be found again quickly and easily. Few things are more discouraging than to be unable to locate an article that you found once and know will be helpful if you could locate it again.

It is usually a good procedure to record each reference on a card or sheet of paper or maintain a computer database. English teachers usually recommend the use of file cards, but engineers

TA 152.17
G273

Inganere, M.E.
Heat, Air, and Gas Power. McGraw-Hill, New York, 1973.

Has good tables in appendix and formulas on pages 52 - 55 covering heat transfer cases that may occur on project.

Figure 3.8
Documentation of research findings is essential if the findings are to be useful later. Recording pertinent information on a card or computer database will permit easy retrieval of the findings when needed.

seem to prefer information put in a bound notebook or the computer. Whatever your choice, Fig. 3.8 is recommended as a reasonable form of record.

As we are looking at something like a piece of equipment, we often have thoughts and ideas that should be recorded for future reference. At such times our ability to sketch is an invaluable tool because so many details can be graphically shown but are very difficult to describe in words.

3.4.4
The Chapter Example—Step 3

The team of engineering students whose project we are following realized the importance of the research phase but were also aware of the overall time constraints on the design process. They decided that a detailed research plan was needed so that specific assignments could be made to avoid conflict or overlap of effort. After considerable discussion specific research areas were assigned to each team member. They consulted home builders about the demand for fireplaces, suppliers of firewood, several manufacturers of chain saws, companies that sell chain saws, a landscape architect, city government, the library, a county extension service, an engineer, and a company that sells commercial log-splitters.

One of the team members was assigned the task of checking the library to determine what products were currently available. The librarian explained the various indexes, so the student selected the *Thomas Register*. After failing to find anything listed under "Logs," he tried "Splitters." Figure 3.9 is a reproduction of his notes. (The appendix of the student report includes copies of letters sent by the design team along with the responses received.)

With this information, the student went to the *Yellow Pages* of the local telephone book, wherein he learned that one of the products was sold locally; so a team member was assigned to visit the dealer. The dealer was temporarily out of advertising pamphlets, so the team member sketched the floor model and recorded pertinent data about it, for which the notes are shown as Fig. 3.10. Their procedure is to be commended and highly recommended in that they adequately documented the information in sufficient detail so that it could be used or the manufacturer could be contacted for additional data.

It must be noted here that this design project was performed prior to the availability of the WWW at this university. Today, of course, the team members would be searching for data on the Web.

The initial research stage has provided us with added information about the problem so that we are now ready to begin to describe the design in terms of things it must be or must have and what attributes are most important.

Source: Thomas Register of American Manufacturers
Thomas Register Catalog File

Listing: Splitters: Wood, Firewood, Kindling, etc.

Location: Engineering Library—Reference Tables

<u>Manufacturers</u>

1. Gordon Corporation
   P.O. Box 244-TR
   Farmington, CT
   (Hydraulic)

2. H. L. Diehl Co.
   South Windham, CT

3. Vermeer Mfg. Co.
   3804 New Sharon Rd.
   Pella, IA
   (Powered, log, hydraulic, trailer)
   (515) 628-3141

4. Tree King Mfg. & Engineering, Inc.
   North St.
   Showhegan, ME
   (Hydraulic)

5. Lindig Mfg. Corp.
   1831 West Country Rd.
   St. Paul, MN
   (612) 633-3072

6. Carthage Machine Co., Inc.
   571 West 3rd Ave.
   Carthage, NY
   (Wood for pulp mills)

7. Equipment Design & Fabrication, Inc.
   722 N. Smith St.
   Charlotte, NC
   (Log)

8. Pabco Fluid Power Co.
   5752 Hillside Ave.
   Cincinnati, OH
   (Log)
   (513) 941-6200

9. Piqua Engineering, Inc.
   234-52 First Street
   Piqua, OH
   (513) 773-2464

10. Hencke Mfg. Co., Inc
    433 W. Florida St.
    Milwaukee, WI
    (Log)

11. Didier Mfg. Co.
    1652 Philips Ave.
    P.O. Box 806
    Racine, WI
    (Hydraulic, log)

12. Murray Machinery, Inc.
    104 Murray Road
    Wausau, WI
    (Hydraulic, paper roll)

Figure 3.9
A record of manufacturers who
produce log-splitters.

3.5
Constraints—Step 4

Up to this point we have kept the problem definition as broad as possible, allowing for a maximum number of potential solutions. However, there are physical and practical limitations (called constraints) that will reduce the number of solutions for any problem. For example, in order to remain competitive in the marketplace, a retail cost of $50 for a particular solution cannot be exceeded. Another example is that any conceived solution must be able to operate using a standard household 120-V (volt) outlet. In many cases the physical size of the solution is limited by market competition or legal restrictions. The laptop computer is a good example of placing size (and weight) constraints on a solution in order to meet the competition in the marketplace. Whenever a *constraint* is applied, the solution possibilities are reduced; therefore, you should take care that the constraint is not "artificial," that is, not overly restrictive to the process of finding an innovative solution.

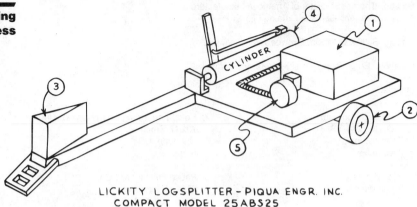

LICKITY LOGSPLITTER – PIQUA ENGR. INC.
COMPACT MODEL 25ABS25

OVERALL SPECS
LENGTH   74.5"
WIDTH   32.5"
HEIGHT   23.0"(W/CONTROL LEVER
         FOLDED)

① 5 H.P. BRIGGS & STRATTON
② 8" SEMI-PNEUMATIC WHEELS
③ HEAT TREATED WEDGE
④ RAM FORCE – 10 TONS
⑤ HYDRAULIC PUMP

Figure 3.10
A sketch is an excellent method
of recording certain types of
information.

We face a similar situation in almost every decision we make, even though sometimes they are not really important. When you arose this morning, you had to choose some clothes to wear. You probably limited the choice to those hanging in your closet (or maybe in your roommate's closet). This was a constraint *you* placed on the situation, not one that really existed until you made it so. In most fields of engineering formulas have been developed and are used in designs of various kinds. Many, and probably most, of them are valid in a certain range of physical conditions. For instance, the hydraulic conditions of the flow of water are not valid below 0 °C or above 100 °C and are restricted to normal pressure ranges. We normally refer to these constraints or limits as *boundary conditions,* and they occur in many different ways.

3.5.1
The Chapter Example—Step 4

The student design team did not list any specific constraints, but we must assume that they are practical individuals who would assign a very low rating to a concept that exceeded some level of performance. For instance, if one of their ideas had an estimated cost of $500, it would no doubt be rated at near zero. They did not, however, tell us at what cost the zero rating begins; is it $200 or $100 or $75? In like manner, they did not say at what weight they would consider a concept to be too heavy to be portable. They did give us a little help by restricting the projected users to adults (excluding children).

Criteria are desirable characteristics of the solution which are established from experience, research, market studies, and customer preferences. In most instances the criteria are used to judge alternative solutions on a qualitative basis. However, if a performance function can be defined mathematically, then an optimum solution can be obtained numerically. The mathematical method of optimization is beyond the scope of this introductory text. Instead we concentrate on the selection and weighting of a set of criteria that will produce the best solution for the stated problem and solution constraints.

3.6.1
Design Criteria

Whereas each project or problem has a personality all its own, there are certain characteristics that occur in one form or another in a great many projects. We should ask ourselves, "What characteristics are most desirable and which are not applicable?" Typical design criteria are listed below:

1. Cost—almost always a heavily weighted factor
2. Reliability
3. Weight (either light or heavy)
4. Ease of operation and maintenance
5. Appearance
6. Compatibility
7. Safety features
8. Noise level
9. Effectiveness
10. Durability
11. Feasibility
12. Acceptance

There will be other criteria, and perhaps some of those given are of little or no importance in some projects, so a design team in industry or in the classroom must decide which *criteria* are important to the design effort (see Fig. 3.11). Since value judgments have to be made later, it probably makes little sense to include those criteria that will be given relatively low weights. There are often mild disagreements at this point, not about which criteria are valid but, rather, about how much weight should be assigned to each. It is frequently better if the team members make their assignments of weight independently and then compile all the results. This tends to dampen the effect of the more persuasive members at the same time that it forces all team members to contribute consciously. Usually, there are not many instances where one of the members strongly disagrees

Figure 3.11
The simulation of accidents on a computer system assists engineers in the establishment of design criteria for safer vehicles. *(Courtesy of the ISU Visualization Laboratory.)*

with the mean value of the weight assigned to each criterion. Some negotiation may be required, but it is seldom a difficult situation to resolve.

3.6.2
The Chapter Example—Step 5

Most people feel comfortable when they talk in general terms about a great many things because as long as they do not have to get specific, an avenue of escape from their position is left open. The students whose progress we are following were not so fortunate for they were facing a real problem. A review of their time-line schedule indicated that it was time to make some decision. They therefore agreed upon the following criteria and assigned weights:

1. Cost: 30 percent
2. Portability: 20 percent
3. Ease of operation: 15 percent
4. Safety: 15 percent
5. Durability: 10 percent
6. Use of standard parts: 10 percent

What areas of agreement and disagreement do you see between our list of 12 criteria and their list of 6?

The most obvious difference is that they included the use of standard parts as an important criterion, but it was not listed at all in the section on Design Criteria. Just why this is important is not clear unless you try to place yourself in their position. If the team has plans to manufacture the log-splitter, then it will be much easier and less expensive to begin operations if many components can be purchased rather than manufactured in their own plant.

They agreed that cost, weight (portability), ease of operation, and safety were important. The others on our list of 12 were either not considered or considered to be of low importance (less than 5 percent).

It is our conclusion that at this point the students have decided that the solution to their problem will be a portable log-splitter that can be operated by a single adult. If this is true, then we can conclude that a loop has been installed in the design process. This is not unusual. Figure 3.4 might very well have a number of arrows to show that problem solvers do return to steps in the design process that have supposedly been completed. In this particular case we can assume that our students have redefined their problem even though they do not say so and do not report having undertaken any additional research. Again, this is not unusual.

3.7
Alternative Solutions—Step 6

Suppose that you are chief engineer for a manufacturing company and are faced with appointing someone to the position of director of product testing. This is an important position because all the company's products are given rigorous testing under this person's direction before they are approved. You must compare all the candidates with the job description (criteria) to see who would do the best job. This seems to be a ridiculously simple procedure, doesn't it? Well, we think it does too, but many times such a process is not followed and poor appointments are made. In the same way that the list of candidates for the position has to be made, so must we produce a list of possible answers to our problem before we can go about the job of selecting the best one.

3.7.1
The Nature of Invention

The word "invention" strikes fear into the minds of many people. They say, "Me, an inventor?" The answer is "Why not?" One reason that we do not fashion ourselves as inventors is that some of our earliest teaching directed us to be like other children. Since much of our learning was by watching others, we learned how to conform. We also learned that if we were like the other kids, no one laughed at us. We can recall in our early days, even preschool, that the worst thing that could happen would be if people laughed at us. We will bet that most of you have a similar feeling when you say something that is not too astute, and it is followed by smiles and polite laughs. Moreover, we do not like to experiment because many experiments fail. It is a very secure person who never has to try something that he or she has not already done well. Think about it: When you were in the first few grades at school, didn't you feel great when you were called on by the teacher and you knew the answer? Don't you, even today, try to avoid asking your professors a question because you do not want the professor or your classmates to know that you do not know the answer? Most of us like to be in the majority. Please do not assume that we are saying that the majority of the people are wrong or that it is bad to be like other people. However, if we dwell on such behavior, then we will never do anything new. A degree of inventiveness or creativity is essential if we are to arrive at solutions to problems that are better than the way things are being done now. If we can remove the blocks to creativity, then we have a good chance of being inventive.

The father of one of the authors had a motto above his desk that read as follows:

> Life's greatest art,
> Learned through its hardest knocks,
> Is to make
> Stepping stones of stumbling blocks.

He does not know whether or not this was original with his father, but he remembers it after not having seen it for over 20 years, and it surely applies to the process of developing ideas.

3.7.2
Building the List

There are a great number of techniques that can be used to assist us in developing a list of possible solutions. Three of the more effective methods will be briefly discussed.

Checkoff lists, designed to direct your thinking, have been developed by a number of people. Generally, the lists suggest possible ways that an existing solution to your problem might be changed and used. Can it be made a different color, a different shape, stronger or weaker, larger or smaller, longer or shorter, of a different material, reversed or combined with something else? It is suggested that you write your list down on paper and try to conceive of how the current solution to the problem might be if you changed it according to each of the words on your list. Ask yourself: Why is it like it is; will change make it better or worse; did the original designers have good reason for doing what they did or did they simply follow the lead of their predecessors?

Consider, as an application of checkoff lists, the log-splitter solution shown in Fig. 3.10. Use the checklist words "modify" and "rearrange" to guide or focus our efforts to obtain new solutions.

modify?	Use electric power instead of gas.
	Mount in the box of a pickup truck.
rearrange?	Vertical cylinder action (easier to handle logs).
	Wedge moved by cylinder (not fixed as shown).

Morphological listing gives a visual conception of the possible combinations that might be generated. These listings are usually shown as grids or diagrams. It is easy to visualize a rectangular prism as shown in Fig. 3.12. This example indicates that we are

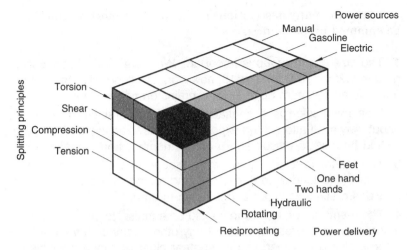

Figure 3.12
A morphological chart. There are 72 combinations of the three alternatives. The shaded segment, for example, would employ an electric reciprocating device that splits by torsion.

considering the log-splitting problem as composed of three subdivisions: power source, power delivery, and splitting principles. These are then subdivided, as shown in Fig. 3.12. The prism produces 72 different combinations such as the one indicated by the shaded volume. Here we would have the logs split by torsion applied by a reciprocating motion generated by an electric power source. Surely you can think of more than the three major subdivisions shown in Fig. 3.12 and can provide additional ideas for each of them.

Brainstorming is a technique that has received wide discussion and support. The mechanics of a brainstorming session are rather simple. The leader states the problem clearly and ideas about its solution are invited. The length of productive sessions varies, but it is usually in the half-hour range. Often it takes a few minutes for a group to rid itself of its natural reserved attitude. But brainstorming can be fun, so choose a problem area and try it with some friends. Be prepared for a surprise at the number of ideas that we will develop (see Fig. 3.13).

Figure 3.13
A student design team brainstorms for ideas to detect and resist motion in two directions, a requirement for their design of a stepper-climber exercise machine.

There are many descriptions of this process, most of which can be summed up as follows:

1. The size of the group is important. We have read of successful groups that range from 3 to 15; however, it is generally agreed that 6 to 8 is an optimum number for a brainstorming session.

2. Free expression is essential. That is what brainstorming is all about. Any evaluation of the exposed ideas is to be avoided. Nothing should be said to discourage a group member from speaking out.

3. The leader is a key figure, even though free expression is the hallmark. The leader sets the tone and tempo of the session and provides a stimulus when things begin to drag.

4. The members of the group should be equals. No one should feel any reason to impress or support any other member. If your supervisor is also a member, you must steer clear of concern for his or her feelings or support for his or her ideas.

5. Recorders are necessary. Everything that is said should be recorded, mechanically or manually. Evaluation comes later.

We have discussed a few techniques that are recommended to stimulate our thought processes. You may choose one of the free-wheeling techniques or perhaps a well-defined method. Regardless of your preferences, we think you will be pleased and even surprised at the large list of ideas that you can develop in a short period of time.

3.7.3
The Chapter Example—Step 6

Our team of engineering students approached their task of generating ideas by setting up a brainstorming session. Their minds were already tuned in on the problem; they had produced a list of candidates for the ultimate decision. They admit that they erred by giving preliminary evaluation of some of the ideas, which is a hard rule not to violate. The following is a list of ideas as it appeared in their written report. It is not, however, their total list:

1. Hydraulic cylinder (vertical or horizontal) used as a method to apply force.

2. Auto-jack or fence-tightener concept in order to apply pressure through a mechanical advantage (see Fig. 3.14).

3. Use of compressed air to force the wedge through the log.

4. Adaptations of conventional hand tools such as the axe, mall, or wedge.

5. Power or manual saws.

6. Heavy pile driver with a block and tackle for raising the weight.

7. High-voltage arc between electrodes; similar to a lightning bolt.

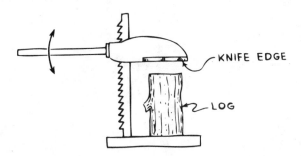

KNIFE EDGE

LOG

Figure 3.14
Thumbnail sketches are often
helpful in describing ideas.

8. Spring-powered wedge using either compression or tension.

9. Sliding mass that drives the wedge into the wood.

10. Drop the wedge from the elevated position on the log.

11. Electronic sound that produces compression waves strong enough to split logs.

12. Wedge driven by explosive charge.

13. Spinning hammermill that breaks by shearing and concussion like a rock crusher.

14. Separate or split with intensive concentrated high energy such as a laser beam.

15. Force a conical wedge into a log and apply a torsional force.

16. Use a large mechanical vice with one jaw acting as a wedge.

17. Drill core (hole) in wood, fill with water, cap, and freeze.

18. Cut wood into slabs rather than across the grain.

19. Apply a force couple to the ends of a log, causing a shearing action.

20. Drop a log from an elevated position onto a fixed wedge.

3.8
Analysis—Step 7

At this point in the design process we have defined the problem, expanded our knowledge of the problem with a concentrated search for information, established constraints for a solution, selected criteria for comparing solutions, and generated alternative solutions. In order to determine the best solution in light of available knowledge and criteria, the alternative solutions must be analyzed to determine performance capability. Analysis thus becomes a pivotal point in the design process. Potential solutions which do not prove out during the analysis phase may be discarded or, under certain conditions, may be retained with a redefinition of the problem and a change in constraints or criteria. Thus, one may need to repeat segments of the design process (Fig. 3.4) after completing the analysis.

Analysis involves the use of mathematical and engineering principles to determine the performance of a solution. Consider a system, such as the cantilever beam in Fig. 3.15a, constrained by the laws of nature. When there is an input to the system (the applied load P), analysis will determine the performance of the

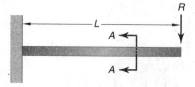

Figure 3.15 (a)

	P, N	L, m	h,m	b,m	E, Pa	I, m^4	d,m
	1.00E+05	4	0.1	0.2	2.07E+11	1.67E-05	0.618357
	1.00E+05	4	0.2	0.2	2.07E+11	0.000133	0.077295
	1.00E+05	4	0.3	0.2	2.07E+11	0.00045	0.022902
	1.00E+05	4	0.4	0.2	2.07E+11	0.001067	0.009662
	1.00E+05	4	0.5	0.2	2.07E+11	0.002083	0.004947
	1.00E+05	4	0.6	0.2	2.07E+11	0.0036	0.002863
CANTILEVER BEAM DEFLECTION FOR RECTANGULAR SECTION							

Figure 3.15 (b)

system (beam)—that is, the deflection, stress buildup, etc. Keep
in mind that the objective of design is to determine the best solution (system) to a need.

The following example should be studied for the process rather
than the content. The goal is to obtain quantitative information
for the decision step in the design process.

Example problem 3.1 Determine the deflection of the
beam in Fig. 3.15 under the following conditions. Assume the beam
is structural steel.

$L = 4.0$ m

$h = 0.40$ m

$b = 0.20$ m

$P = 1.0 \times 10^5$ N

Solution The deflection of the end of a cantilever beam for the
configuration shown is given by

$$d = \frac{PL^3}{3EI} \quad \text{(constraint equation)}$$

where

d = deflection, m

E = modulus of elasticity, a material constant, Pa

$= 2.07(10^{11})$ Pa for structural steel

I = moment of inertia, m^4

For a rectangular cross section

$$I = \frac{bh^3}{12}$$

$$= \frac{(0.2)(0.4)^3}{12}$$

$$= 1.067(10^{-3})\text{m}^4$$

Therefore,

$$d = \frac{(10^5)(4)^3}{3(2.07)(10^{11})(1.067)(10^{-3})}$$

$$= 9.66(10^{-3})\text{m}$$

$$= 9.7 \text{ mm}$$

This result would be forwarded to the designer for incorporation into the decision phase. The time required to produce an analysis is critical to the design process. If it takes longer to do an analysis than the schedule (Fig. 3.5) permits, then the results are somewhat meaningless. The engineer must exercise some judgment in selecting the method of analysis in order to assure results within the time limit. You can visualize the potential of computers in the analysis effort. Many alternatives can be investigated in a brief amount of time. It is a simple task to change any of the parameters—b, h, P, L, or E—and to see the effect on the deflection immediately. Fig. 3.15b shows a spreadsheet printout of cantilever beam deflections for a rectangular section for varying depths of the section while holding all other values constant. It would have been possible to produce plots of each parameter vs. deflection by using computer graphics. It is obvious that the more possibilities one can investigate, the better the problem is understood and the better the design will be.

We will discuss the beam of Example prob. 3.1 in more detail in Sec. 3.8 to illustrate further the role of analysis in engineering design.

Analysis performed by engineers in most design projects is based on the laws of nature, the laws of economics, and common sense.

3.8.1
The Laws of Nature

You have already come into contact with many of the laws of nature, and you will no doubt be exposed to many more. At this point in your education you may have been exposed to the conservation principles: the conservation of mass, of energy, of momentum, and of charge. From chemistry you are familiar with the laws of Charles, Boyle, and Gay-Lussac. In mechanics of materials Hooke's law is a statement of the relationship between load and deformation. Newton's three principles serve as the basis of analysis of forces and the resulting motion and reactions.

Many methods exist to test the validity of an idea against the laws of nature. We might test the validity of an idea by constructing a mathematical model, for example. A good model will allow us to vary one parameter many times and to examine the behavior of the other parameters. We may very well determine the limits within which we can work. Other times we will find that our boundary conditions have been violated, and therefore the idea must be modified or discarded.

Results of an analysis of a mathematical model are frequently presented as graphs. Often the slopes of tangents to curves, points of intersection of curves, areas under or over or between curves, or other characteristics provide us with data that can be used directly in our designs.

Figure 3.16
Advanced solid modeling software is capable of generating very complex models such as that of this human hand. Note the tendons and muscle fiber.

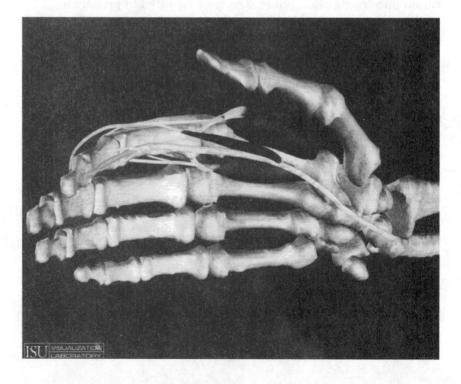

Figure 3.17
Students obtain a plot of data directly as the experiment progresses. The experiment can quickly be adjusted or redone if the plot does not follow theoretical predictions.

Computer graphics enables a mathematical model to be displayed on a screen. As parameters are varied, the changes in the model and its performance can also be quickly displayed to the engineer.

The preparation of scale models of proposed designs is often a necessary step. This can be a simple cardboard cutout or it can involve the expenditure of great sums of money to test the model under simulated conditions that will predict how the real thing will perform under actual use. A prototype or pilot plant is sometimes justified because the cost of a failure is too great to chance. Such a decision usually comes only after other less expensive alternatives have been shown to be inadequate.

You probably have surmised that the more time and money that you allot to your model, the more reliable are the data that you receive. This fact is often distressing because we want and need good data but have to balance our needs against the available time and money.

3.8.2
The Laws of Economics

Section 3.8.1 introduced the idea that money and economics are part of engineering design and decision making. We live in a society that is based on economics and competition. It is no doubt true that many good ideas never get tried because they are deemed to be economically infeasible. Most of us have been aware of this condition in our daily lives. We started with our parents explaining why we could not have some item that we wanted because it cost too much. Likewise, we will not put some very desirable component into our designs because the value gained will not return enough profit in relation to its cost.

Industry is continually looking for new products of all types. Some are desired because the current product is not competing well in the marketplace. Others are tried simply because it appears that people will buy them. How do manufacturers know that a new product will be popular? They seldom know with certainty. Statistics is an important consideration in market analysis. Most of you will find that probability and statistics are an integral part of your chosen engineering curriculum. The techniques of this area of mathematics allow us to make inferences about how large groups of people will react based on the reactions of a few. It is beyond our study at this time to discuss the techniques, but industry routinely employs such studies and invests millions of dollars based on the results.

3.8.3
Common Sense

We must never allow ourselves the luxury of failing to check our work. We must also judge the reasonableness of the result.

During the 1930s, the depression years, a national magazine conducted a survey of voters and predicted a Republican victory. The magazine was wrong, and the public lost confidence in it to the point that the magazine went out of business. The editors had sampled the population by taking all the telephone books in the United States and, by a system of random numbers, selecting people to be called. They then applied good statistical analysis and made their prediction. Why did they miss so badly? It is a bit hard for us today to imagine the depression years, but the facts are that large percentages of voters did not have telephones, so this economic class of people was not included in the analysis. But this group of people did vote, and they voted largely Democratic. Nothing was wrong with the analytical method, only the basic premise. The message is rather obvious: No matter how advanced our mathematical analysis, the results cannot be better than our basic assumptions. Likewise, we must always test our answers to see if they are reasonable.

3.8.4
The Chapter Example—Step 7

Our design team generated many ideas for splitting wood, 20 of which are listed in Sec. 3.7.3. In addition, the criteria had been previously determined (see Sec. 3.6.2) and value decisions made with regard to evaluating the importance of each criterion. At this point decisions had to be made to reduce the number of alternative solutions. The time available for completing the design project did not allow the team the luxury of thorough analysis of each of the ideas. Therefore, a decision was made to reduce the number of alternative solutions to five. These five were then investigated and developed in more detail. The following is the result of the team's analysis.

Analysis of Alternative Solutions

(Items marked with asterisks were kept by the team for further development.)

1. Hydraulic cylinder (vertical or horizontal)
   a. Extreme cost for materials and manufacturing
   b. High operational and maintenance costs
   c. Nonportable for one person
   d. Lack of standard parts

*2. Auto-jack principle or fence tightener (force by creating a mechanical advantage)
   a. Reasonably portable
   b. Minimum manual labor required

3. Use of compressed air (pneumatic)
   a. Minimum portability
   b. Extensive material, manufacturing, and operational costs

4. Adaptations of conventional hand tools such as the axe, mall, or wedge
   a. Inefficient operation
   b. Is the current solution
   c. Unsafe for an inexperienced user

5. Power or manual saws
   a. High cost of materials and manufacturing
   b. Not a low-volume solution

6. Heavy pile driver with block and tackle used to raise weight
   a. Not portable
   b. Expensive

7. High-voltage arc between electrodes, similar to lightning bolt
   a. Inefficient
   b. Expensive
   c. Impractical to use

*8. Spring-powered wedge using either compression or tension
   a. Relatively easy to use
   b. Portable
   c. Low manufacturing cost

*9. Sliding mass that drives a wedge into wood
   a. Good for low-volume usage
   b. Portable
   c. Low initial cost and operational costs

*10. Drop a wedge from an elevated position onto the log
  a. Uses a mechanical advantage
  b. Simple construction
  c. Good for low-volume production

11. Electronically produced sound that produces compression waves strong enough to split logs
  a. Impractical because of other damage that could be done
  b. Dangerous for average person to use

*12. Wedge driven by explosive charge
  a. Minimum work
  b. Low cost
  c. Not easily portable

13. Hammermill that would chop wood much like a coal or rock crusher
  a. Expensive
  b. Much waste material (chips)
  c. Not easily portable

14. Separate or split with concentrated high energy from a laser beam
  a. Expensive
  b. Potentially dangerous

15. Force a conical wedge into log and apply a torsional force
  a. Complicated mechanical design for a single piece of equipment
  b. Expensive

16. Use a large vise with one jaw acting as a wedge
  a. Slow operating if powered by human
  b. Probably not easily portable

17. Drill core (hole) in wood, fill with water, cap, and freeze
  a. Time-consuming
  b. Inefficient

18. Cut wood into slabs rather than splitting
  a. Inefficient
  b. Not suitable for ordinary fireplace

19. Apply a force couple to the ends of a log, causing a shearing action
  a. Expensive
  b. Inefficient
  c. Destructive to wood

20. Drop a log from an elevated position onto a fixed wedge
  a. High amount of manual labor required
  b. Inefficient

You will note that the analysis is based solely on the stated criteria and at best is very general. Many of the comments made show that no computations involving the mechanics of wood splitting were made. No testing (or test data) was used for any of the potential solutions. [A professional engineer would almost surely have spent considerable time and money on models, simulations, prototypes, and tests to verify (or disprove) his or her ideas about the many aspects of the concepts before attempting a decision.] You should decide for yourself whether the team should be criticized at this point in their design effort. Remember, the team had only a few weeks together and limited experience in engineering analysis courses.

We will list at this time a few analyses that may have been made by beginning engineering students. You may agree or disagree with some of them and perhaps add other items.

1. Determine the force necessary to split a log of a given type of wood. This may be done experimentally or analytically. Some data on this exist in the literature.

2. Determine the stamina of a human being with regard to lifting a specified weight a number of times in a given time period. This would be valuable information for the manually operated splitters.

3. For impact-type splitters, make a preliminary estimation of impulse required to split logs. That is, ascertain what combinations of mass and velocity are needed to cleave the log.

4. Find out if there is a wedge angle that would be more efficient than others.

5. Decide what masses we are talking about with respect to the alternative solutions. Since portability is a major consideration, locate or generate some data to be able to make comparisons based on the total mass of the various devices.

We will illustrate the analysis necessary to estimate the sliding mass that would drive a wedge into wood (alternative solution 9 given above).

Team members who are just beginning their study of engineering may find the following analysis difficult to understand completely. A better understanding will come after engineering physics, statics, and dynamics. Another method to obtain a reasonable value for the mass of the wedge is to conduct an experiment using the concept shown. Still another method is to consult an instructor or upper-level engineering student to guide the team through the analysis.

Example problem 3.2   Estimate the mass required for the sliding-mass wood-splitter.

Solution   We assume that the peak force is the critical parameter in causing the wood fibers to separate. We know from experience that logs can be split with a wedge driven by a sledge hammer (see Fig. 3.18a). We will attempt to generate the same impact

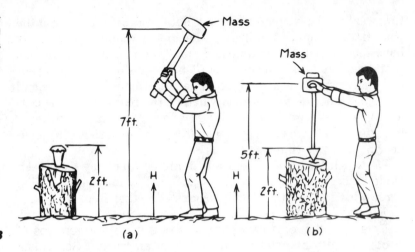

Figure 3.18   (a)   (b)

momentum with the sliding mass as with the sledge. The configuration for the sliding mass is shown in Fig. 3.18b. The problem will be illustrated using British units since that is what the design team used for their work.

From physics

Change in momentum = impulse

$$\Delta mV = F\,\Delta t \text{ for a constant force}$$

Energy change = work done

$$\Delta KE + \Delta PE = F\,\Delta H \text{ for a constant force}$$

where m = mass, slugs

V = velocity, ft/s

F = force, lbf

t = time, s

KE = kinetic energy, lbm $\cdot$ ft^2/s^2

$\quad = \frac{1}{2}mV^2$

PE = potential energy, lbm $\cdot$ ft^2/s^2

$\quad = mgH$

g = acceleration of gravity, ft/s^2

H = height, ft

Assumptions (refer to Fig. 3.18a):

Mass of sledge = 8/32.2 slug

= 0.2484 slug

Peak height = 7 ft

Average wedge height = 2 ft

Initial velocity = 0

Downward force generated = 10 lbf

Using the subscripts i for initial conditions and f for final conditions, we have for the sledge configuration

$$\Delta KE + \Delta PE = F\,\Delta H$$

$$\frac{1}{2}m(V_f - V_i)^2 + mg(H_f - H_i) = F(H_f - H_i)$$

$$\frac{1}{2}mV_f^2 = F(H_f - H_i) + mg(H_i - H_f)$$

$$\frac{1}{2}(0 \cdot 248\ 4)\,V_f^2 = (-10)\ (2 - 7)$$
$$+\ (0.248\ 4)(32.2)(7 - 2)$$

$$V_f^2 = \frac{50 + 40}{0.1242}$$

$$= 724.64$$

$$V_f = 26.92 \text{ ft/s}$$

Thus, the change in momentum of the sledge is

$$\Delta mV = m(V_f - V_i)$$

$$= 0.248\ 4(26.92 - 0)$$

$$= 6.687 \text{ slug} \cdot \text{ft/s}$$

Now with the subscript s representing the sliding mass, we make the following assumptions (refer to Fig. 3.18b):

Peak height = 5 ft

Average wedge height = 2 ft

Downward force generated = 10 lbf

$$V_{is} = 0$$

Therefore, for the sliding mass configuration

$$\Delta KE + \Delta PE = F\Delta H$$

$$\frac{1}{2}m_s(V_{fs} - V_{is})^2 + m_s g(H_{fs} - H_{is}) = F_s(H_{fs} - H_{is})$$

For the same impact momentum

$$m_s V_{fs} = 6.687 \text{ slug} \cdot \text{ft/s}$$

$$\frac{1}{2}m_s V_{fs}^2 = m_s(32.2)(5 - 2) + (-10)(2 - 5)$$

$$\frac{m_s}{2}\left(\frac{6.687}{m_s}\right)^2 = 96.6m_s + 30$$

$$22.36 = 96.6m_s^2 + 30m_s$$

$$96.6m_s^2 + 30m_s - 22.36 = 0$$

Solving for m_s and discarding the unrealistic answer, we obtain

$$m_s = 0.350 \text{ slug}$$

$$= 11.3 \text{ lbm}$$

3.9

Decision—Step 8

As we mentally review our own professional practices, we can honestly say that the most difficult times for us have not been when the analysis of a problem was difficult but, rather, when it required a "tough" decision. We have known many engineers who are technically knowledgeable but are unable to make a final decision. They may be happy to suggest several possible solutions and to outline the strong and weak points of each—indeed they may feel that their function is to do just that—but let someone else decide which course is to be followed. The truth of the matter is that most engineering assignments require both: providing information and making decisions.

What makes reaching a decision so difficult? The answer is *trade-offs*. If we can be certain about anything in the future, it is that with your decisions, the necessity to compromise will come. Review the criteria in Sec. 3.5.2 that our team selected. In order to sell the log-splitter that they are to design, it must be competitive in cost—so the lower the cost, the better. If the engineers wish to make it more durable, chances are that they will use materials that are expensive or use more material (heavier construction perhaps). If they use more material, they are adding weight, which limits portability.

Each time one criterion is optimized, another moves away from its optimum position. If the relationships are complicated, you may have to go through very complex processes to reach a decision. You can be certain that no one idea will be better in all respects than all the others; hence, you may have to choose a concept that you know is inferior to others in one or more of the decision criteria.

3.9.1
Organization for Decision

In order to decide among several alternatives, you need as much information as possible about each. In design you need information in order to evaluate each alternative against each of the criteria. Analysis can provide the answers, as described in Sec. 3.7. If time and money are available, a prototype may be constructed and tested. In most cases judgment must be made with much less information. Computer models and engineering drawings are used to describe the form and function of the design.

Whatever information is available, it should fairly and accurately represent all the alternatives so that an equitable decision can be made.

3.9.2
The Chapter Example—Step 8

The analysis process (see Sec. 3.7.4) reduced our student team's list of ideas to five, which are repeated here for reference.

1. Auto-jack principle—the pressure wedge (item 2)
2. Spring-powered wedge (item 8)
3. Sliding mass (item 9)
4. Wedge dropped from an elevation (item 10)
5. Wedge driven by explosive charge (item 12)

Each team member was assigned the difficult and creative task of taking one of the five very general ideas and beginning the process of shaping it into physical form. Again, the student design team was limited in time.

A solution or physical shape for any of the five ideas listed can assume an infinite number of different forms. Two of the five solutions are illustrated in this chapter: idea 1, the pressure wedge, and idea 3, the sliding mass. The other three areas were also developed but are not illustrated here.

Each student took one possible solution and prepared a series of idea sketches. An idea sketch can be a single view, a multiview, or a pictorial. The sketches include little detail but clearly depict the form and function of the idea. Figure 3.19 illustrates three of the better idea sketches created for the pressure wedge idea.

A second student took the sliding-mass idea and developed the three separate sliding configurations illustrated in Fig. 3.20.

Obviously, the end product will be a refinement of this effort, so as many idea sketches as possible should be developed in the available time. When no additional time remains for analysis, it is necessary to evaluate each idea sketch. This phase involves considerable detail and is often referred to as "concept development." The final concepts must be developed to the point that comparative judgments can be intelligently made in evaluating each concept in light of the criteria.

Concept sketches are shown for both the pressure wedge (Fig. 3.21) and the sliding mass (Fig. 3.22).

3.9.3
Criteria in Decision

The objective of the entire design process is to choose the best solution for a problem within the time allowed. The steps that pre-

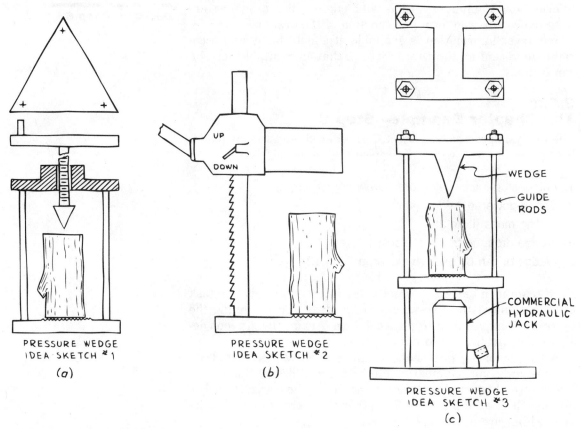

Figure 3.19
Three different idea sketches for the pressure wedge.

PRESSURE WEDGE
IDEA SKETCH #1

(a)

PRESSURE WEDGE
IDEA SKETCH #2

(b)

WEDGE

GUIDE
RODS

COMMERCIAL
HYDRAULIC
JACK

PRESSURE WEDGE
IDEA SKETCH #3

(c)

cede the decision phase are designed to give information that leads to the best decision. It should be quite obvious by now that poor research, a less-than-adequate list of alternatives, or inept analysis would reduce one's chances of selecting a good, much less the best, solution. Decision making, like engineering itself, is both an art and a science. There have been significant changes during the past few decades that have changed decision making from being primarily an art to what it is at the present, with probability, statistics, optimization, and utility theory all routinely used. It is not our purpose to explore these topics, but simply to note their influence and to consider for a moment our task of selecting the best of the proposed solutions to our problem. The term "optimization" is almost self-explanatory in that it emphasizes that what we seek is the best, or optimum, value in light of a criterion. As you study more mathematics, you will acquire more powerful tools through calculus and numerical methods for optimization.

In order to illustrate optimization, we will return to the beam problem illustrated in Fig. 3.15 and Example prob. 3.1. Our objective will be to determine the least mass of the beam for prescribed performance conditions. You will recall in our discussion

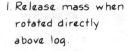

1. Release mass when rotated directly above log.
2. Rotate and reset mass.

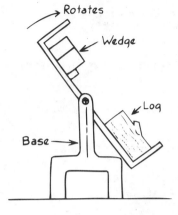

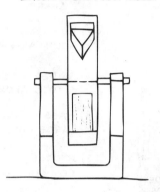

SLIDING MASS
IDEA SKETCH #1

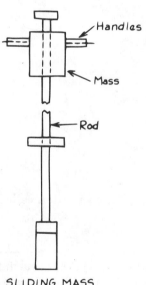

SLIDING MASS
IDEA SKETCH #2

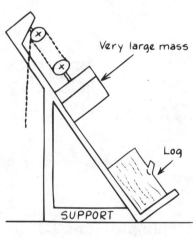

SLIDING MASS
IDEA SKETCH #3

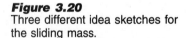

Figure 3.20
Three different idea sketches for the sliding mass.

of analysis that a system, the laws of nature and economics, an input to the system, and an output are involved. Analysis gives the output if the system, laws, and input are known.

If we consider the inverse problem—that is, if we were looking for a system, given the laws, input, and output—we would be using *synthesis* rather than analysis. Synthesis is not as straightforward as analysis since it is possible to have more than one system that will perform as desired. But if we specify a criterion for selecting the best solution, then a unique solution is possible.

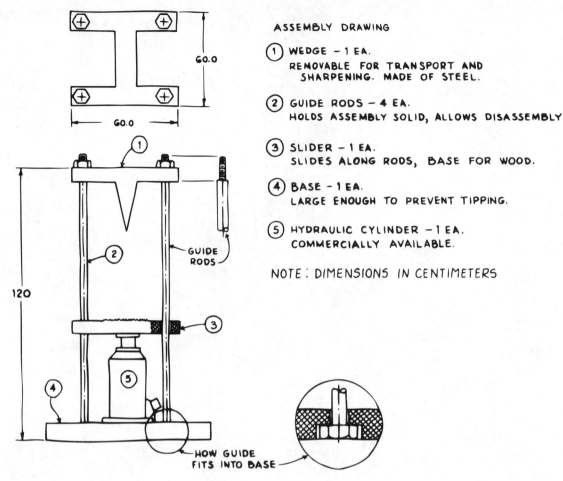

ASSEMBLY DRAWING

1 WEDGE – 1 EA.
   REMOVABLE FOR TRANSPORT AND
   SHARPENING. MADE OF STEEL.

2 GUIDE RODS – 4 EA.
   HOLDS ASSEMBLY SOLID, ALLOWS DISASSEMBLY

3 SLIDER – 1 EA.
   SLIDES ALONG RODS, BASE FOR WOOD.

4 BASE – 1 EA.
   LARGE ENOUGH TO PREVENT TIPPING.

5 HYDRAULIC CYLINDER – 1 EA.
   COMMERCIALLY AVAILABLE.

NOTE: DIMENSIONS IN CENTIMETERS

60.0

60.0

120

GUIDE
RODS

HOW GUIDE
FITS INTO BASE

Figure 3.21
Concept development of the
pressure-wedge idea, sketch no.
3 from Fig. 3.19.

The criterion used for selecting the best solution is often called
a payoff function.

Example problem 3.3 (Refer to Fig. 3.15.) Determine the
dimensions b, h for the least beam mass under the following conditions:

The deflection cannot exceed 40 mm.

The height h cannot be greater than three times the base b.

$E = 2.0 \times 10^{11}$ Pa

$L = 4.0$ m

$P = 1.0 \times 10^5$ N

If the beam has a constant cross section throughout, then the
mass is a minimum when the cross-sectional area $A = bh$ is a minimum. Achieving minimum mass by finding minimum area (payoff
function) will provide the best (optimum) solution.

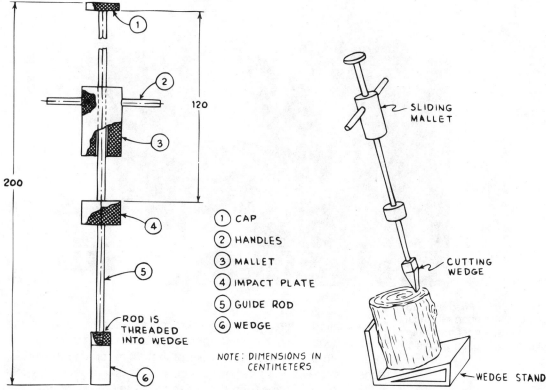

① CAP
② HANDLES
③ MALLET
④ IMPACT PLATE
⑤ GUIDE ROD
⑥ WEDGE

NOTE: DIMENSIONS IN CENTIMETERS

ROD IS THREADED INTO WEDGE

SLIDING MALLET

CUTTING WEDGE

WEDGE STAND

Figure 3.22
Concept development for the sliding-mass idea, sketch no. 2 from Fig. 3.20. This concept was selected as the best solution by the team.

Solution The system we are after is the beam shape $b \times h$ within the conditions specified above; the law is the deflection equation from Example prob. 3.1; the inputs are L, P, and E, and the output is the range of permissible deflection. The deflection equation becomes

$$d = \frac{PL^3}{3EI}$$

$$0.04 = \frac{10^5(4)^3}{3(2)(10^{11})I}$$

or

$$I = 2.667(10^{-4})\text{m}^4$$

Then

$$\frac{bh^3}{12} = 2.667(10^{-4})$$

Thus,

$$b = \frac{3.2(10^{-3})}{h^3}$$

Figure 3.23
Two members of an electrical engineering design team ponder results of a computer analysis.

This equation is a relationship for the beam under the condition that the deflection is a constant 40 mm. The expression is plotted in Fig. 3.24a. Note that values to the right of the curve represent beam dimensions for which the deflection would be less than 40 mm. Those to the left would cause the deflection to exceed 40 mm; thus, that portion of the *design space* for b and h is invalid.

Next we demonstrate the effect of the required relationship between b and h by plotting the line b = h/3, as shown in Fig. 3.24a. Points above this line represent valid geometric configurations; those below do not.

Now we have a better picture of the design space, or solution region for our problem. A point h = 0.3 m, b = 0.3 m, represents a satisfactory solution since it falls within all conditions *except* possibly minimum mass. Many designs stop at this point when a nominal solution has been found. These are the designs that may not survive in the marketplace because they are not optimum. In fact, to get a nominal solution, we could have guessed values for b and h and very quickly had an answer without going to all the effort we have up to this point.

We know the region in which the best solution (minimum area) lies. We again take advantage of the capability of the spreadsheet and have it search for the solution we desire. Figure 3.24b shows a table of values for h and the two constraints on the solution, namely a deflection of 40 mm or less and that h cannot be greater than three times b. We set up the spreadsheet to iterate on the payoff function, A = bh, within the constraints, and find a minimum. The solution, in this case, is the intersection of the two curves in Fig. 3.24a. Often, we find that optimum solutions lie on the boundary

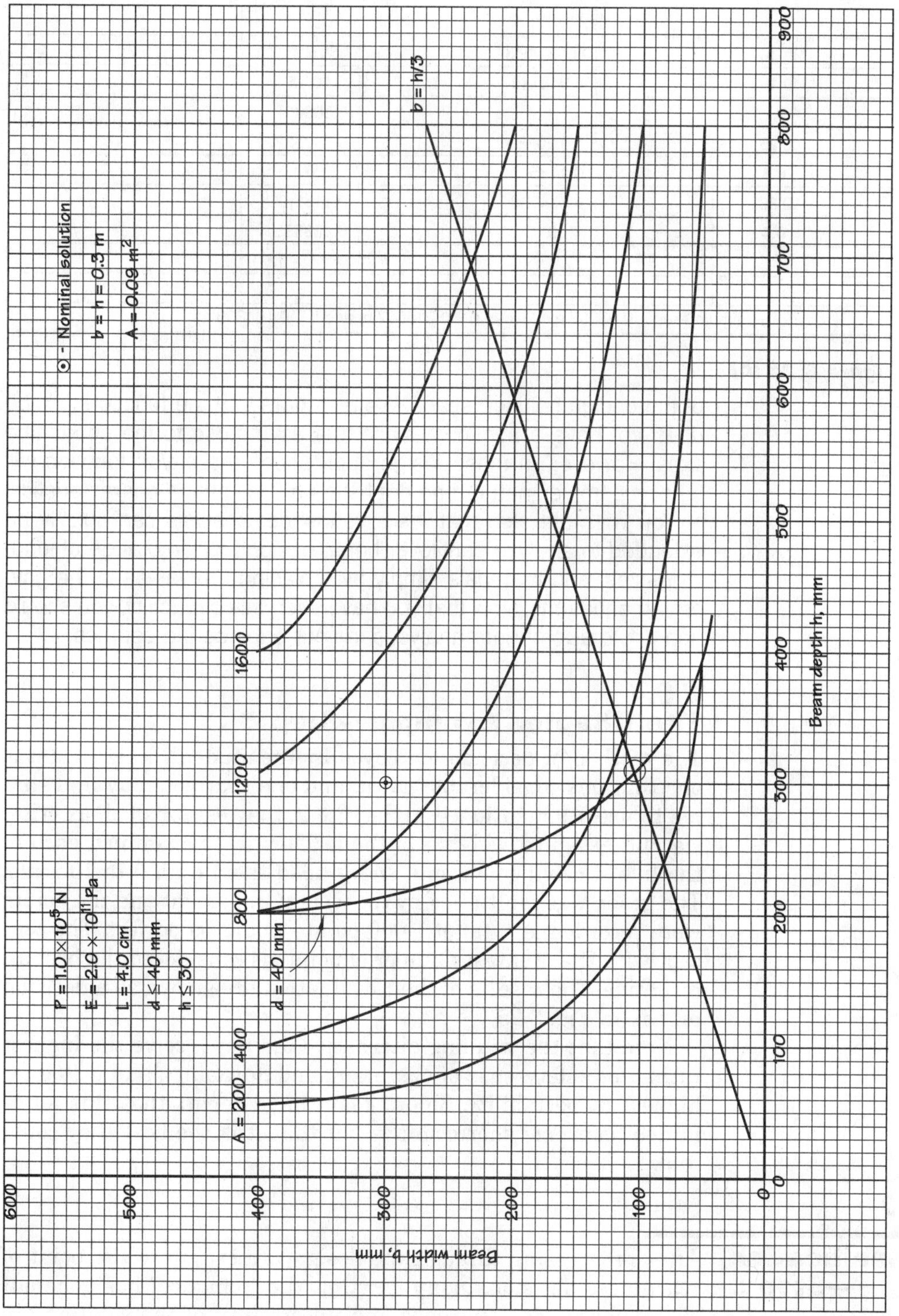
⊙ - Nominal solution
b = h = 0.3 m
A = 0.09 m²
P = 1.0 × 10⁵ N
E = 2.0 × 10¹¹ Pa
L = 4.0 cm
d ≤ 40 mm
h ≤ 30
d = 40 mm
b = h/3
A = 200
400
800
1200
1600
Beam depth h, mm
Beam width b, mm
0
100
200
300
400
500
600
700
800
900

Figure 3.24 (a)

	h, m		0.0032/h^3		b=h/3	
	0.15		0.948148		0.05	
	0.2		0.4		0.066667	
	0.25		0.2048		0.083333	
	0.3		0.118519		0.1	
	0.35		0.074636		0.116667	
	0.4		0.05		0.133333	
	Minimum Area = 0.03266 m^2 when h = 0.313017 m and b = 0.104339 m					

Figure 3.24 (b)

of the design solution space. We also note that the function $A = bh$ does not have a minimum since it is asymptotic to the coordinate axes. However, in most design applications the payoff function will be of a complex nature and not have obvious shape characteristics.

You will note also that the spreadsheet analysis can quickly be extended to analyze cantilever beam designs for any range of loading, sizes, and materials simply by replacing the values in the governing equations. Now that you have seen this capability, you should apply it to appropriate analyses in your engineering courses.

3.9.4
The Chapter Example—Step 8

The method of decision employed by the student design team is one that has considerable merit, and one that you may find simple to use. As mentioned in Sec. 3.6.2, they established six criteria and assigned weights to them. They later examined each of the five surviving concepts and graded them on a 0 to 10 scale. The grade was multiplied by the weight in percent and the points were recorded. A total of 1 000 points would be perfect. The concept with the greatest number of points is considered the best alternative. The results of their evaluation are shown as a decision matrix in Fig. 3.25.

Note that the winning alternative did not receive the highest rating for safety, ease of operation, and durability, and it tied for highest in portability. So our team must report that this alternative has some shortcomings but that it is the best they could find.

3.10

Specification —Step 9

After progressing through the design process up to the point of reaching a decision, many feel that the "romance" has gone out of the object. The suspense and uncertainty of the solution are over, but much work still lies ahead. Even if the new idea is not a breakthrough in technology but simply an improvement in existing technology, it must be clearly defined to others. Many very

Decision matrix

Criteria	Weight W, percent	Selected concepts (see below) 1	2	3	4	5	6
Cost	30	6 / 180	7 / 210	7 / 210	7 / 210	9 / 270	
Ease of operation	20	10 / 200	7 / 140	9 / 180	10 / 200	7 / 140	
Safety	15	9 / 135	7 / 105	6 / 90	5 / 75	8 / 120	
Portability	15	6 / 90	5 / 75	4 / 60	10 / 150	10 / 150	
Durability	10	8 / 80	9 / 90	10 / 100	9 / 90	9 / 90	
Use of standard parts	10	7 / 70	8 / 80	8 / 80	6 / 60	9 / 90	
Total	100	775	700	720	785	860	

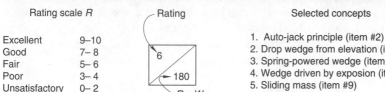

Rating scale R	
Excellent	9–10
Good	7–8
Fair	5–6
Poor	3–4
Unsatisfactory	0–2

Selected concepts

1. Auto-jack principle (item #2)
2. Drop wedge from elevation (item #10)
3. Spring-powered wedge (item #8)
4. Wedge driven by exposion (item #12)
5. Sliding mass (item #9)
6. Additional concepts

Figure 3.25
Each concept was rated by the team on a scale of 0 to 10 for each criterion. The rating was multiplied by the criterion weight and then summed. Concept 5 was chosen as the optimum even though it did not receive the highest rating for three of the six criteria.

creative people are ill-equipped to convey to others just exactly what their proposed solution is. It is not the time to use vague generalities about the general scope and approximate size and shape of the chosen concept. One must be extremely specific about all the details regardless of the apparent minor role that each may play in the finished product.

3.10.1
Specification by Words

One medium of communication that the successful engineer must master is that of language, written and spoken. Your problems in professional practice would be considerably less complicated if you were required to defend and explain your ideas only to other engineers. Few engineers have such luxury; most must be able to write and speak clearly and concisely to people who do not have comparable technical competence and experience. They may be officials of government who are bound so tightly by budgets and the need for public acceptance that only the best explanation is good enough to pierce their protective armor. They may be people in business who know that capital is limited and that they cannot defend another poor report to stockholders. Without appropriate documentation, they will not be nearly as certain as you are that your idea is a good one. Therefore, this phase of specification—*communication*—is so important that we have assigned it as the final step in the design process. Before discussing it, however, we will discuss another means of communication.

3.10.2
Graphical Specifications

You will probably have many occasions in which to work closely with technicians and drafters as they prepare the countless drawings that are essential to the manufacture of your design. You will not be able to do your job properly if you cannot sketch well enough to portray your idea or to read drawings well enough to know whether the plans that you must approve will actually result in your idea being constructed as you desire.

A lathe operator in the shop, an electronics technician, a contractor, or someone else must produce your design. How is the person to know what the finished product is to look like, what materials are to be used, what thicknesses are required, how it is to work, what clearances and tolerances are demanded, how it is to be assembled, how it is to be taken apart for maintenance and repair, what fasteners are to be used, and so on?

Typical drawings that are normally required include:

1. A sufficient number of detail drawings describing the size and shape of each part
2. Layouts to delineate clearances and operational characteristics
3. Assembly and subassemblies to clarify the relationship of parts
4. Written notes, standards, specifications, etc., concerning quality and tolerances
5. A complete bill of materials

Included with the drawings are almost always written specifications, although certain classes of engineering work refer simply to documented standard specifications. For example, most cities have adopted one of several national building codes, so all structures constructed in those cities must conform to the code. It is quite common for engineers and architects to refer simply to the building code as part of the written specifications and to write detailed specifications for only items that are not covered in the code. This procedure saves time and money for all by providing uniformity in bidding procedures. Many groups have produced

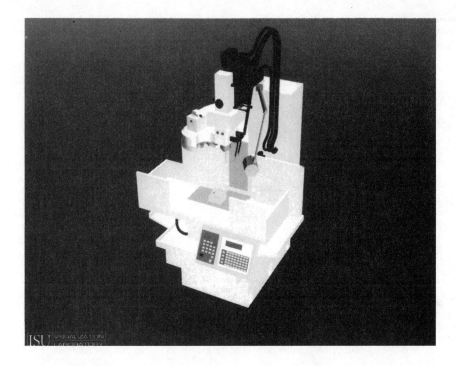

Figure 3.26
Manufacturing capability can be modeled and simulated on the computer as in this example of a milling machine.

standards that are widely recognized. For instance, there are standards for welds and fasteners for the obvious reasons—ease of specification and economy of manufacture. Moreover, there are such standards for each discipline of engineering.

3.10.3
The Chapter Example—Step 9

Reproductions of two drawings prepared by the student team are given as Figs. 3.27 and 3.28. The following drawings were part of their completed report:

1. Detailed drawings of all eight parts of the log-splitter (one shown)
2. Exploded pictorial of the log-splitter
3. Detail drawing of the wedge stand (see Fig. 3.22 for pictorial)
4. Welding assembly of the log-splitter
5. Welding assembly of the wedge stand
6. Complete parts list

The drawings were accompanied by a cost analysis, weight summary, and description of the operating characteristics.

None of the team worked in a plant that produces such items, so no doubt an experienced detailer or drafter would find reason to be critical of their drawings, but we feel that this phase of the process was performed quite well and that there will be few misunderstandings as a result of omissions on their part.

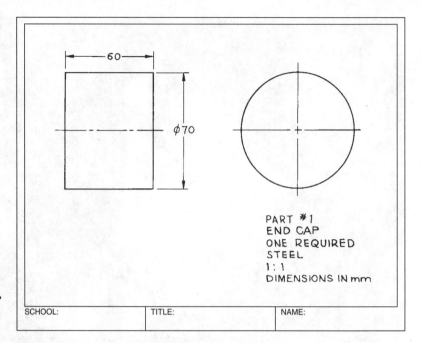

| SCHOOL: | TITLE: | NAME: |

Figure 3.27
Detail drawing of part #1 (see Fig. 3.28).

3.11

Communication —Step 10

During the 1960s the word "communication" seemed to take on a very high priority at conferences and in the professional journals. The need for conveying information and ideas had not changed, but there was an awareness of too much incomplete and inaccurate rendering of information. At most of the professional conferences the authors have attended, one or more papers either discussed the need for engineers to develop greater skills in communication or demonstrated a technique for improving the skills. Students at most universities are required to complete freshman English courses and, in some colleges, a technical writing course in their junior or senior year; but many professors and employers feel that not enough emphasis is being placed on the application of communication skills.

For our purposes here, however, we will discuss only the salient points involved in design step 10.

3.11.1
Selling the Design

It is certainly the responsibility of any profession to inform people of findings and developments. Engineering is no exception in this regard. Our emphasis here will be on a second type of communication, however: selling, explaining, and persuading.

Selling takes place all the way through the design process. Individuals who are the most skillful at it will see many of their

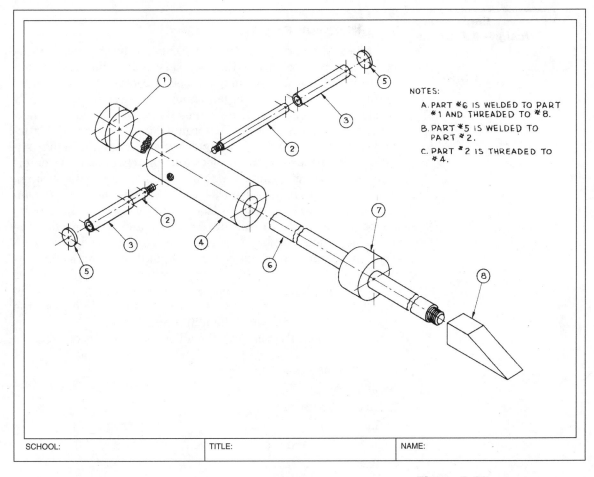

NOTES:

A. PART #6 IS WELDED TO PART #1 AND THREADED TO #8.

B. PART #5 IS WELDED TO PART #2.

C. PART #2 IS THREADED TO #4.

SCHOOL: TITLE: NAME:

Figure 3.28
Exploded pictorial drawing of the log-splitter.

ideas develop into realities. Those who are not so good at it will no doubt become frustrated with their supervisors for not exploring what they feel is a perfectly good idea in more depth.

If you are working as a design engineer for an industry, you cannot simply decide on your own that you will try to improve the product line. Industry is anxious for its engineers to initiate ideas, but will not necessarily approve all of them. As an engineer with a company, you must convince those who decide what assignments you get that the idea is worth the time and money required to develop it. Later, after the design has developed to the point where it can be produced and tested, you must again persuade management to place it into production.

It is a natural reaction to feel that your design has so many clear advantages that selling it should not be necessary. Such may be the case; but in actual situations, things seldom work so smoothly and simply. You will be selling or persuading or convincing others almost daily in a variety of ways. Among the many forms of communication are written and oral reports.

117

3.11.2
The Written Report

The types of reports that you will write as an engineer will be varied, so a precise outline that will serve for all the reports cannot be supplied. The two major types of reports are those used by individuals within the organization and those used primarily by clients or customers. Many times the in-house reports follow a strict form prescribed by the organization, whereas those intended for the client are usually designed for the particular situation. The nature of the project and the client usually determine the degree of formality employed in the report. Clients often state that they wish to use the report in some particular manner, which may direct you to the style of report to use. For instance, if you are a consultant for a city and have studied the needs for expansion of a power plant, the report may be very technical, brief, and full of equations, computations, and so on, if it is intended for the use of the city's engineer and public utilities director. However, if the report is to be presented to citizens in an effort to convince them to vote for a bond issue to finance the expansion, the report will take a different flavor.

Reports generally will have the following divisions or sections:

1. Appropriate cover page
2. Abstract
3. Table of contents
4. Body
5. Conclusions and recommendations
6. Appendixes

Abstract. A brief paragraph indicating the purpose and results of the effort being reported, the abstract is used primarily for archiving so that others can quickly decide if they want to obtain the complete report.

Body. This is the principle section of the report. It begins with an introduction to activities, including problem identification, background material, and the plan for attacking and solving the problem. If tests were conducted, research completed, and surveys undertaken, the results are recounted and their significance is underscored. In essence, the body of the report is the description of the individual or team effort on a project.

Conclusions and recommendations. This section tells why the study was done and explains the purpose of the report. Herein you explain what you now believe to be true as a result of the work discussed in the body and what you recommend should be done about it. You must lay the groundwork earlier in the report, and at this point you must sell your idea. If you have done the job carefully and fully, you may make a sale; but do not be discouraged if you do not. There will be other days and other projects.

Appendixes. Appendixes can be used to avoid interrupting your descriptions so that it can flow more smoothly. Those who do not want to know everything about your study can read it without digression. What is in the appendix completes the story by showing all that was done. But it should not contain information that is essential to one's understanding of the report.

It should be emphasized that all reports do not follow a specific format. For example, lengthy reports should have a summary section placed near the beginning. This one- or two-page section should include a brief statement of the problem, the proposed solution, the anticipated costs, and the benefits. The summary is for the use of higher-level management who in general do not have the time to read your entire report.

In many instances your instructor or supervisor will have specific requirements for a report. Each report is designed to accomplish a specific goal.

Student reports often must follow an instructor's directions regarding form and topics that must be included. If you are asked to write a report in an introductory design course, refer to Fig. 3.1 to make sure that all of the steps in the design process have been successfully completed. You might also study the bias of your instructor and make sure that you have done especially the steps that he or she considers most important. This may sound as though pleasing the instructor and getting a good grade is all that is important. But perhaps in this respect the academic situation is something like that in industry or private practice: Your report must take into account the audience—its biases and its expectations, whether professor in the classroom or supervisor in the business world.

3.11.3
The Oral Presentation

The objective of the oral presentation is the same as that of the written report—to furnish information and to convince the listener. However, the methods and techniques are quite different. The written report is designed to be glanced at, read, and then studied. The oral presentation is a one-shot deal that must be done quickly, so it must be simple. There is no time to go into detail, to show complicated graphs and tables of data or many of the things that are given in a written report. What can you do to make a good presentation?

First, you must be prepared. No audience listens to people who have not bothered to prepare themselves. So you should rehearse with a timer, a mirror, and a tape recorder.

Stand in such a way that you do not detract from what you are saying or showing.

Look at your audience and maintain eye contact. You will be receiving cues from those who are listening, so be prepared to react to these cues.

Project your voice by consciously speaking to the back row. The audience quickly loses interest if it has to struggle to hear.

Speak clearly. We all have problems with our voices—they are either too high, too low, or too accented and certain words or sounds are hard for us, but always be concerned for the listener.

Preparation obviously includes being thoroughly familiar with the material. It should also include determining the nature, size, and technical competence of the audience. You must know how much time will be allotted to your presentation and what else, if anything, is to be presented before or after your speech. It is essential that you know what the room is like because the physical conditions of the room—its size, lighting, acoustics, and seating arrangements—may very well control your use of slides, transparencies, video, and microphones.

The quality of your graphic displays can often influence the opinion of your audience. Again, consider to whom you are speaking carefully as you choose which and how many displays to use. Be certain that they can be read and understood or do not use them at all. Do not clutter your displays with so many details that the message is obscured. Do not try to make a single visual aid accomplish too many tasks: It is good to change the center of emphasis. By all means, test your visual aids before the meeting and never apologize for their quality. (If they are not good, do not use them.) Figure 3.29 shows an exploded pictorial of a small battery-powered grinding tool and provides a good overview of the tool's components.

Figure 3.30 narrows the focus to a subassembly of the tool. The quality of your visual aids can influence many people for you or against you before they hear all you have to say.

Have a good finish. Save something important for the last and make sure everyone knows when the end has come. By all means, do not end with "Well, I guess that's about all I have to say." You have much more to say, you just do not have the time to say it.

3.11.4
The Chapter Example—Step 10

The written and oral reports presented by the students were significant parts of their design experience. Both reports were regulated somewhat by their professor in much the same way that reports are in industry. They were told who would be reading the report and who would judge the oral presentation. They were given copies of the written report grading sheet and the oral pre-

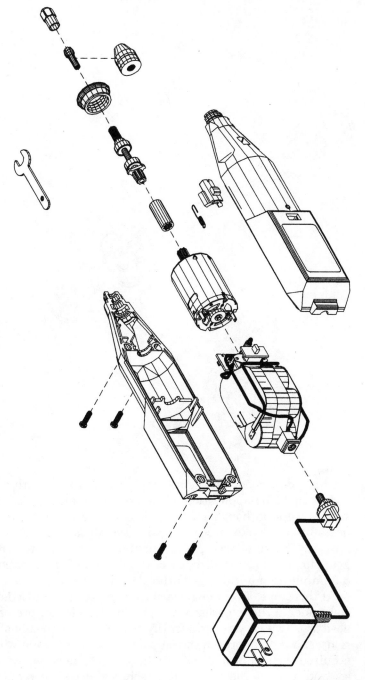

Figure 3.29
Exploded pictorial shows all
parts of the grinding tool.

sentation judging card. They correctly accepted these constraints
as real (not imagined), and they were given high ratings by their
evaluators.

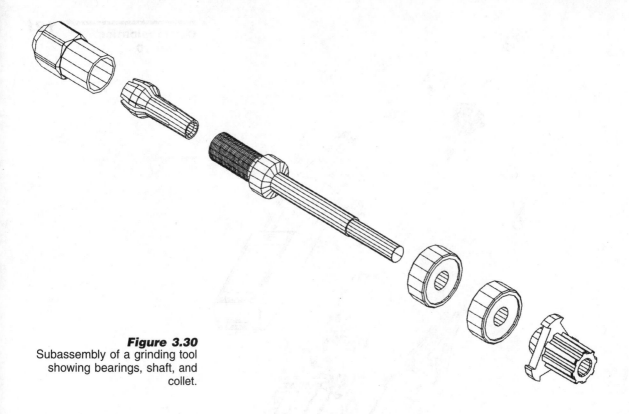

Figure 3.30
Subassembly of a grinding tool showing bearings, shaft, and collet.

3.12

Key Terms and Concepts

We have described the engineering design process with a series of 10 steps. The application of the process to an engineering design process, although structured, is an iterative process with flexibility to make necessary adjustments as the design progresses. The emphasis in this chapter is on conceptual design. You should realize that many of the details of specification and communication of the log-splitter design are not included.

At this stage of your engineering education it is important that you undergo the experience of applying the design process to a need with which you can identify based on your personal experiences. As you approach the baccalaureate degree, you will have acquired the technical capability to conduct the necessary analyses and to make the appropriate technical decisions required for complex products, systems, and processes.

Some of the terms and concepts you should recognize and understand are listed below:

Engineering design   Constraint
Design process   Criteria

Reverse engineering	Analysis
Alternative solutions	Decision matrix
Checkoff lists	Synthesis
Morphological listing	Payoff function
Brainstorming	Specification
Design space	Communication

Problems

Problems 3.1 and 3.2 involve synthesis that is similar to that illustrated in Example prob. 3.3.

3.1 For beam configuration of Fig. 3.15 determine the dimensions b, h for the least mass under the following conditions:

The deflection cannot exceed 50 mm.

The height h cannot be greater than b.

$E = 2.0 \times 10^{11}$ Pa

$L = 6.0$ m

$P = 1.0 \times 10^5$ N

Produce a brief report containing a computer plot that is similar to Fig. 3.14a and a discussion of the design space and how the solution was found.

3.2 A company transfers packages from point to point across the country. The limit on package size is that the girth plus the longest dimension (measured on the package) cannot exceed 60 in. Consider two kinds of packages, a rectangular-prism shape with square ends and a cylindrical shape where the cylinder height is greater than the diameter (girth = circumference). Determine for each shape the largest package volume that can be shipped and the package dimensions at the maximum volume. A *suggested procedure* follows:

    (a) Write the constraint equation (60-in limit) and the payoff function (volume). Eliminate one of the unknowns in the payoff function by substituting the constraint equation.

    (b) Plot the payoff function against the remaining variable.

    (c) Determine the optimum values.

    (d) Prepare a report of your findings.

3.3 Investigate current designs for one or more common items. If you do not have the items in your possession, purchase them or borrow from friends. Conduct the following "reverse engineering" procedures on each of the items:

    (a) Write down the need that the design satisfies.

    (b) Disassemble the item and list all the parts by name.

    (c) Write down the function of each of the parts in the item.

    (d) Reassemble the item.

    (e) Write down answers to the following questions:

- Does the item satisfactorily solve the need you stated in part *a*?

- What are the strengths of the design?

- What are the weaknesses of the design?

- Can this design be easily modified to solve other needs? If so, what needs and what modifications need to be made?

- What other designs can solve the stated need?

The items for your study are the following:

- Mechanical pencil
- Safety razors from three vendors; include one disposable razor
- Flashlight
- Battery-powered slide viewer
- Battery-powered fabric shaver

3.4 The following list of potential projects can be approached in the manner used by the student design team featured in this chapter.

- Headlights that follow the wheels' direction
- A protective "garage" that can be stored in the car's trunk
- A device to prevent theft of helmets left on motorcycles
- A conversion kit for winter operation of motorcycles
- An improved rack for carrying packages or books on a motorcycle or bicycle
- A child's seat for a motorcycle or bicycle
- A tray for eating, writing, and playing games in the back seat of a car
- A system for improving traction on ice without studs or chains
- An inexpensive built-in jack for raising a car
- An auto-engine warmer
- A better way of informing motorists of speed limits, road conditions, hazards, etc.
- Theft- and vibration-proof wheel covers
- A better way to check the engine oil level
- A device to permit easier draining of the oil pan by weekend mechanics
- A heated steering wheel for cold weather
- A less expensive replacement for auto air-cleaner elements
- An overdrive system for a trail bike
- A sun shield for an automobile
- A well-engineered, efficient automobile instrument panel
- An SOS sign for cars stalled on freeways
- A remote car-starting system for warmup
- A car-door positioner for windy days
- A bicycle trailer
- Automatic rate-sensitive windshield wipers
- A corn detasseler
- An improved wall outlet
- A beverage holder for a card table
- A car wash for pickups
- A better rural mailbox
- A home safe
- An improved automobile traffic pattern on campus
- An alert for drowsy or sleeping drivers

- An improved automobile headlight
- An improved bicycle for recreation or racing
- Improved bicycle brakes
- A transit system for campus
- A pleasure boat with retractable trailer wheels
- Improved pedestrian crossings at busy intersections
- A transportation system within large airports
- An improved baggage-handling system at airports
- Improved parking facilities in and around campus
- A simple but effective device to assist in cleaning clogged drains
- A device to attach to a paint can for pouring
- An improved soap dispenser
- A better method of locking weights to a barbell shaft
- A shoestring fastener to replace the knot
- An automatic moisture-sensitive lawn waterer
- A better harness for seeing-eye dogs
- A better jar opener
- A system or device to improve efficiency of limited closet space
- A shoe transporter and storer
- A pen and pencil holder for college students
- An acceptable rack for mounting electric fans in dormitory windows
- A device to pit fruit without damage
- A riot-quelling device to subdue participants without injury
- An automatic device for selectively admitting and releasing an auxiliary door for pets
- A device to permit a person to open a door when loaded with packages
- A more efficient toothpaste tube
- A fingernail catcher for fingernail clippers
- A more effective alarm clock for reluctant students
- An alarm clock with a display to show it has been set to go off
- A device to help a parent monitor small children's presence and activity in and around the house
- A chair that can rotate, swivel, rock, or stay stationary
- A simple pocket alarm that is difficult to shut off, for discouraging muggers
- An improved storage system for luggage, books, etc. in dormitories
- A lampshade designed to permit one to study while his or her roommate is asleep
- A device that would permit blind people to vote in an otherwise conventional voting booth
- A one-cup coffeemaker
- A solar greenhouse
- A quick-connect garden-hose coupling

- A device for recycling household water
- A silent wakeup alarm
- Home aids for the blind (or deaf)
- A safer, more efficient, and quieter air mover for room use
- A lock that can be opened by a secret method without a key
- A can crusher
- A rain-sensitive house window
- A better grass catcher for a riding lawnmower
- A winch for hunters of large game
- Gauges for water, transmission fluid, etc. in autos
- A built-in auto refrigerator
- A better camp cooler
- A dormitory-room cooler
- A device for raising and lowering TV racks in the classroom
- An impact hammer adapter for electric drills
- An improved method of detecting and controlling the level position of the bucket on a bucket loader
- Shields to prevent corn spillage where the drag line dumps into the sheller elevator (angle varies)
- An automatic tractor-trailer-hitch aligning device
- A jack designed expressly for motorcycle use (special problems involved)
- A motorbike using available (junk) materials
- Improved road signs for speed limits, curves, deer crossings, etc.
- More effective windshield wipers
- A windshield de-icer
- Shock-absorbing bumpers for minor accidents
- A home fire-alarm device
- A means of evacuating buildings in case of fire
- Automatic light switches for rooms
- A carbon monoxide detector
- An indicator to report the need for an oil change
- A collector for dust (smoke) particles from stacks
- A means of disposing of or recycling soft-drink containers
- A way to stop dust storms, resultant soil loss, and air entrainment
- An attractive system for handling trash on campus
- A self-decaying disposable container
- A device for dealing with oil slicks
- A means of preventing heat loss from greenhouses
- A way of creating energy from waste
- A bookshelf with horizontally and vertically adjustable shelves and dividers
- A device that would make the working surface of graphics desks adjustable in height and slope, retaining the existing top and pedestal

- An egg container (light, strong, compact) for camping and canoeing
- Ramps or other facilities for handicapped students
- A multifunctional (suitcase/chair/bookshelf, etc.) packing device for students
- A self-sharpening pencil for drafting
- An adapter to provide tilt and elevation control on existing graphics tables
- A compact and inexpensive camp stove for backwoods hiking
- A road trailer operable from inside the car
- A hood lock for cars to prevent vandalism
- A system to prevent car thefts
- A keyless lock

Index